In the Circle of Ancient Trees

Edited by
VALERIE TROUET

Art by **Blaze Cyan**

DAVID SUZUKI INSTITUTE

GREYSTONE BOOKS
Vancouver/Berkeley/London

Published in North America by Greystone Books in 2025
Published in the United Kingdom,
copyright © UniPress Books Limited 2025
Illustrations copyright © 2025 by Blaze Cyan

25 26 27 28 29   5 4 3 2 1

Greystone Books Ltd.
greystonebooks.com

David Suzuki Institute
davidsuzukiinstitute.org

Cataloguing data available from Library and Archives Canada
ISBN 978-1-77840-268-5 (cloth)
ISBN 978-1-77840-269-2 (epub)

Cover and text design by Wayne Blades
Printed and bound in China on FSC® certified paper. The FSC® label
means that materials used for the product have been responsibly sourced.

Greystone Books thanks the Canada Council for the Arts, the British Columbia
Arts Council, the Province of British Columbia through the Book Publishing Tax Credit,
and the Government of Canada for supporting our publishing activities.

Canadä

Greystone Books gratefully acknowledges the xʷməθkʷəy̓əm (Musqueam),
Sḵwx̱wú7mesh (Squamish), and səlilwətaɬ (Tsleil-Waututh) peoples on
whose land our Vancouver head office is located.

CONTENTS

INTRODUCTION

My mom passed away this summer. She still lived in the big house with the '70s wallpaper in the suburbs of Leuven, a university town in Belgium, where I grew up. Like many people of my age and generation, I spent a couple of months emptying out a house filled to the brim with the collections and memories of two people born before the Second World War, and who had raised three children within its walls. My daunting task and my mourning were lightened by the views of the backyard that I had from every room in the back of the house. In July, I sorted mismatched Tupperware in the kitchen and watched the roses flourish. In August, I organized my dad's book collection in his office as I viewed peaches ripen on the tree outside. In September, I found a supply of diapers for the grandchildren in the upstairs bathroom and gazed at the pheasants exploring the yard. In October, as I finally gathered the courage to tackle the family photo collection in my parents' bedroom, I contemplated the needles on the larch tree turn golden.

I was two years old when my family moved into that house in the mid-1970s. Already then, with its size and ever-changing colors, the larch tree was an impressive presence. By now the tree must be at least eighty years old, a decent age for a backyard tree, and its wide, upward-bending branches still invite pigeons to coo and adult daughters to reminisce. That larch tree is my home tree. It was a witness to my early life and gave me roots, an anchorpoint later on. Its steady presence helped me through my teenage angst and heartbreak, and later, when I moved away from home, first to a dorm, then to another city, then to another country, then to another continent, the tree was always there when I came back. Its melancholic drooping branches always simultaneously asking me, "Where have you been?" and telling me, "Welcome back."

Compared to the millennia-old ancient trees that some of my colleagues work with and whose stories are recorded in this book, the European larch (*Larix decidua*) is rather unremarkable in terms of lifespan. Yet, in other ways, larch trees are fascinating. Larch is the only coniferous species that loses its needles every winter and, as such, it behaves more like its broadleaf, deciduous brethren than its needle-leaf ilk. Larch needles are the only ones you will see turning golden in the fall and lime green in the spring, like you would on an oak or alder. Other conifer trees lose their needles more like humans lose their hair: typically, not all at once, but in bits and pieces over the course of multiple years.

Conifer (or gymnosperm) trees are, evolutionarily speaking, much older than broadleaf (or angiosperm) trees and have a much simpler wood anatomy. Conifer wood is dominated by one type of wood cell, the tracheid, which is square-ish in diameter in cross-section, for instance, when looking at the top of a tree stump through a magnifying glass. Because of their square shape, tracheids are lined up neatly and form straight lines from the most recently formed wood just underneath the bark, to the oldest wood at the tree's pith (or center). Although relatively simple in structure, tracheids are mighty and fulfill two of the three main functions of wood: water transport and mechanical support. The third function, the storage of biochemicals and nutrients, is taken care of by a different, less dominant type of wood cell, the parenchyma cell. In broadleaf trees, the main tasks are divvied up among different, specialized types of wood cells. Water transport from the roots to the leaves in broadleaf trees happens in vessels—large round cells surrounded by smaller fiber cells for mechanical support, and parenchyma cells for storage. In some tree species, such as oak, the vessel cells can be so large in diameter that you can see them clearly with the naked eye.

Regardless of whether they are needle-leaved or broadleaved, trees growing in temperate regions will go into dormancy every winter when days shorten, and temperatures drop. Even in conifer trees that deceivingly retain most of their needles throughout winter, this yearly growth stop leaves its mark in the wood. In spring, when trees are well rested after a long winter sleep and grow vigorously, they form *earlywood*, with large cells that can efficiently transport the water needed to grow a canopy. Conifers form large-diameter tracheids with thin cell walls and broadleaf trees form large vessels with the same purpose. Later in the growing season, once the canopy is formed in late summer and fall, mechanical support takes over from water transport as the primary wood function and the trees form *latewood*. Latewood tracheids in conifers are small in diameter, but have thick, sturdy cell walls. In many broadleaf species, vessels become smaller and fewer as the growing season progresses.

Every year of a temperate tree's life, the tree forms earlywood in spring, latewood in fall, and stops growing in winter. What we see when we see tree rings is the abrupt transition from the small and often darker-colored latewood cells of one year to the large and light-colored earlywood cells of the following year, caused by the yearly growth stop in between the two. In the tropics, where the days are long, warm, and wet enough year-round for trees to grow, the formation of annual tree rings can be complicated, and researching them can be challenging. But even there, in the tropical lowland rainforests, tree species such as cedro (*Cedrela odorata*) or teak (*Tectona grandis*), which form large earlywood vessels and yearly rings, can be found. With a lot of expertise and even more dedication, the stories that these exceptional trees tell can be unearthed.

Thousand-year-old temperate trees have experienced one thousand winters, have stopped and restarted growing one thousand times, and have formed one thousand yearly rings. We can count these rings and measure the width of each of the thousand rings to get an idea of how much the tree grew each year over the past millennium. The first person to realize the scientific potential of the study of tree rings was Andrew Ellicott Douglass (1867–1962), an astronomer who taught at the University of Arizona in Tucson in the early twentieth century. Douglass was interested in solar cycles, eleven-year cycles in the level of activity of the Sun, and he saw tree rings and the information they record as a means to study this interest. By sampling his first twenty-five cross-sections from a log yard in Flagstaff in northern Arizona, Douglass opened a dendrochronological Pandora's box, as it soon turned out that tree rings also record an inconceivable wealth of information in other fields of science. With his strong scientific intuition, Douglass was the first to use tree rings to study the history of forests, climate, and humans. For instance, through his tree-ring research, he was able to attach exact dates to most of the Ancestral Puebloan ruins in the American Southwest. This success, among others, convinced the University of Arizona to let Douglass establish the first Laboratory of Tree-Ring Research in Tucson in 1937. Dendrochronology as a science was born.

To extract tree-ring information from an ancient tree, fortunately, these days we don't need to cut down the tree or even harm it. We use a specialized, hollow increment borer that we drill into the trunk of the tree, parallel to the ground, but at about chest height. We do this by hand, which requires quite a bit of physical strength, especially when coring tree species with hard and dense wood such as oaks or beech, and when coring dozens of trees a day while clambering over boulders on steep terrain. With the increment borer, we can remove a pencil-sized tree core from the tree on which we can immediately see the rings and which contains all the tree-ring information we need. On a lucky day, we do not fall or hurt ourselves en route to the tree we want to core, our borer does not break or get stuck, we do not hit a tree with rot in its center, and the rings that we see on the core are tiny and manifold, so we know that we just cored an ancient tree.

Once back in the lab, we can start crossdating—the process of comparing one sample's tree-ring pattern to that of another. Crossdating is the true nature of dendrochronology, and is at the core (no pun intended) of what makes dendrochronology a science and not just counting rings. Trees that grow in the same region, with the same climate, will experience the same weather changes from year to year, the same dry years and wet years, the same cold summers and hot summers. They will be happy, grow a lot, and form wide rings in the same years and will be unhappy and form narrow rings in the same years. The year-to-year changes in the climate in which they grow will result in common and recognizable patterns of wide and narrow rings in the samples we take from trees from the same region.

For instance, in the semi-arid climate of southern Arizona, the home region of the Laboratory of Tree-Ring Research, water availability is the most important factor determining how much a tree grows each year. Most trees will form narrow rings in dry years and wide rings in wet years. In other, colder regions, such as the European Alps or the boreal forest of Canada and Siberia, the tree-ring pattern will be mostly determined by summer temperatures, rather than water availability. But regardless of whether "good" and "bad" years for tree growth are mostly determined by water availability or by temperature, the sequence of centuries of good and bad climate years is recorded as a recognizable pattern of alternating wide and narrow rings in trees growing in the same region. This is the "Morse code" (wide-wide-narrow-wide-narrow-narrow...) we aim to match between samples when we crossdate.

This code is also preserved in the wood after trees die, after they have been cut down and turned into beams for buildings and planks for ships, or into musical instruments and panels for Renaissance paintings. The pattern is preserved after trees topple over into lakes, rivers, or bogs and their wood is preserved under water, where no oxygen and wood-decaying organisms can reach them. Even in millions-of-years-old petrified wood, in which all organic material has been replaced by mineral deposits, the wood-cell structure and the tree rings are maintained.

The samples we take from living trees with our corer in the field provide the key to cracking the regional Morse code. We know what year we cored the trees and thus the year the most recent ring was formed. By crossdating living tree samples with each other, we can make sure that there are no missing or false rings in our tree-ring sequences, and we can assign an exact year, an absolute date, to each of the hundreds or thousands of rings that we find in ancient, living trees. We can then use these absolutely dated tree-ring samples to build a reference tree-ring chronology that is anchored in the present time by the living trees. Then comes the really cool part: We can crossdate undated samples from dead trees or from archaeological material against the regional reference chronology, find out the year in which the trees died or were cut down, and attach an exact calendar date to each of the rings in the now dated samples. This is how Douglass was able to date when the Ancestral Puebloan ruins in the American Southwest were built. This is how Stradivarius violins and works by the Old Masters who painted on oak panels can be dated and authenticated. And this is how we can extend tree-ring chronologies even farther back in time than the years covered by living ancients.

For instance, the Great Basin bristlecone pines (*Pinus longaeva*) of the American West are the oldest-known living trees on Earth. Some of them have been found to be more than 4,000 years old. The high and dry environment of the Great Basin does not only allow the bristlecones to grow to their extraordinary age, it also provides prime circumstances for preserving the remnant wood of dead trees on the landscape. By crossdating this remnant wood to the bristlecone pine reference chronology, dendrochronologists extended the chronology to cover more than 8,800 years. Even this timespan is exceeded by the German oak chronology, which comprises 6,775 samples from living trees and subfossil

wood found in gravel pits and covers more than 10,500 years without a single gap. Through crossdating with even older subfossil Scots pine (*Pinus sylvestris*) from the same region, the chronology now provides an exact date for each of its rings back to 10,641 BCE (Before Common Era).

I am often asked if I have a favorite tree. When I answer, "The larch tree in my parents' backyard," I am asked if that is the tree that inspired me to dedicate my career to tree rings. Alas, that is not how it went. For girls like me, who grow up in the suburbs of densely populated cities in densely populated countries, inspiring trees are hard to come by. It is not until I moved to the United States and crossdated an incense cedar (*Calocedrus decurrens*) sample from the Sierra Nevada Mountains in California that was more than 1,000 years old that I was inspired. And then again, when I hiked the Pindus mountains in Greece and found myself among 1,000-year-old Bosnian pine (*Pinus heldreichii*) trees. And then again, and then again.

In this book, ten fellow dendrochronologists from all over the world tell the stories of the ten extraordinary tree species that inspired them and that are inseparably intertwined with their professional paths. In our stories, we share our awe for these trees, but also our concern. The trees' perseverance through millennia of trials and tribulations has brought them here, to the present day, where dendrochronologists can read their stories in their rings. Widespread deforestation, as well as our burning of fossil fuels have amped up the threats, the drying, the warming, the wildfires, and the insect outbreaks that test the perseverance of these ancients.

I've been told that books about trees are popular. I hope so. I've also been told that public recognition is growing for the importance of trees for the planet and for our wellbeing. That more and more people share our awe for these phenomenal organisms. I hope so. With our book, we aim to add a time dimension to that awe. We see trees as witnesses. Witnesses of our forest history, of our climate history, and of our human history. But they are also witnesses to our present. Let's give them something good to talk about, two hundred years from now, to future dendrochronologists.

VALERIE TROUET

LABORATORY OF TREE-RING RESEARCH
UNIVERSITY OF ARIZONA
....

1
...

GIANT SEQUOIA

THOMAS W. SWETNAM

SPECIES

Giant sequoia, *Sequoiadendron giganteum*

....

LOCATION

California, USA

....

ESTIMATED AGE

Up to 3,300 years old

The giant sequoia is a tree of superlatives. Its immense size—taller than the Statue of Liberty and about three times its circumference—is truly confounding. The world's largest tree by volume, it towers above the surrounding forest, its trunk of thickened, crimson-tinged bark often reaching over 65 ft (20 m) before its first branches even appear. In fact, these branches alone can be the size of the pines, incense cedars, and firs to be found beneath it.

Its nearly mythical ability to survive is equally astounding. Coming in at a lifespan of sometimes over 3,000 years, its scarred trunk has been known to endure upward of one hundred wildfires in a lifetime.

The giant sequoia is the fire-resistant monarch of the forest, with individual trees often surviving more than one hundred wildfires across their lifespan.

D espite its incredible age, the oldest-known living giant sequoia, at over 3,200 years old, is a relative newcomer when compared to its relatives. At a time when dinosaurs roamed the Earth, its ancestor species once stretched across what is now Canada and America's Pacific Northwest, covering parts of Utah, Nevada, Idaho, central Colorado, and even reaching into western Europe and eastern Asia.

Just as today, some of these ancestor species that grew in western North America as early as the Upper Cretaceous (80–66 million years ago [MYA]) were are also quite large. This continued into the Eocene through to the Pliocene (5.6–2.4 MYA)—an ancient grove at Florissant Fossil Beds National Monument in central Colorado reveals petrified stumps and logs of *Sequoia affinis*. Here, mudflows buried and preserved intact stumps up to 14 ft (4 m) in diameter, along with a remarkably diverse set of tree leaf and needle fossils.

A recent arrival

While its enormous ancestors may have been widespread and around for millennia, the giant sequoia we know today exists naturally in only 75 groves, inhabiting a band of elevation 4,600–7,050 ft (1,400–2,150 m) on the western slopes of the Sierra Nevada mountains, an area of relatively dry and warm summers, and cool, snowy winters. The largest extant groves are about 4,000 acres (1,600 ha) in size and include thousands of large, mature sequoias, while the smallest groves of a few hectares have only dozens of mature trees.

Surprisingly, the giant sequoia is a relatively new arrival to this region of the Sierra Nevada, extending back about 4,500 years. In recent decades, palaeoecological findings from plant parts in packrat middens and pollen from sediment cores show that the locations of giant sequoia groves have shifted since the last ice age (the past 13,000 years). For example, twigs and needles found in packrat middens dating to the last ice age show that sequoias existed at lower elevations than today, though we can't be certain why.

In fact, much of our understanding of sequoia evolution and bio-geography is not clear, largely because preservation of leaf and cone fossils is not as common as petrified tree stems. This lack of leaf and cone fossils at most localities with petrified sequoia-like stems results in uncertainty about how the species diversified, expanded, and contracted. Yet one of the most remarkable findings from pollen and plant parts found in sediment cores from wet meadows is that some modern groves have arrived relatively recently. Earth's oldest-known living giant sequoia, a sapling in 1380 before the Common Era (BCE), may be among only the first or second generation of trees in its grove.

Close neighbors and expanding territory

California's coast redwoods and giant sequoias both have a reddish tinge to their bark and grow to spectacular heights, but they are quite differ-ent in many other respects. The clearest differences are that the coast redwood has a more slender trunk and can grow to greater heights.

As revealed in the name, the coast redwood lives near the Pacific Ocean along an elevational gradient of 100–2,500 ft (30–760 m) above sea level. These are typically foggy and relatively wet places, unlike the Sierra Nevada, where giant sequoias reside. Their needles are also differ-ent. The coast redwood has relatively well-defined flattened needles arrayed along the sides of branchlets, while the giant sequoia has nee-dles compressed in bundle-like arrangements on branchlets, appearing more like juniper leaves than needles typical of most conifers. The cones also differ; coast redwood's cone is small and relatively delicate, 0.5–1 in (1–3 cm) long, whereas the giant sequoia cone is robust and heavier, at 2–3.5 in (5–9 cm) in length.

A good traveler

Besides the coast redwood, the only extant related species is the "dawn redwood" (*Metasequoia glyptostroboides*), discovered in central China in 1943. Yet an overall picture of giant sequoias' range and distribution

GIANT SEQUOIA TIMELINE

• • •

Climatic events and wildfires in giant
sequoia groves in California evidenced by
the tree-ring record

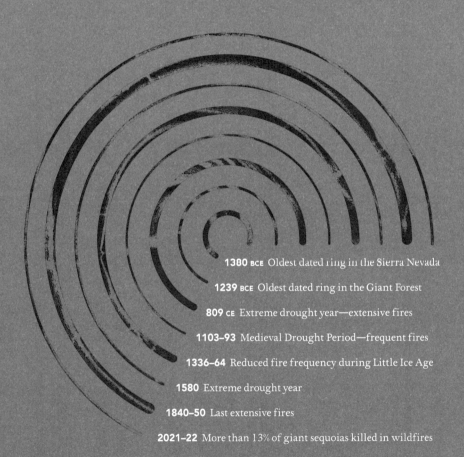

1380 BCE Oldest dated ring in the Sierra Nevada

1239 BCE Oldest dated ring in the Giant Forest

809 CE Extreme drought year—extensive fires

1103–93 Medieval Drought Period—frequent fires

1336–64 Reduced fire frequency during Little Ice Age

1580 Extreme drought year

1840–50 Last extensive fires

2021–22 More than 13% of giant sequoias killed in wildfires

suggests it is a species capable of adapting to climate change, given time and available habitat. An example of its adaptability to growing in different locales is the fact that some of the largest trees today in England are giant sequoias.

Following the "discovery" of giant sequoias and their first scientific naming by a British botanist (naming the genus *Wellingtonia*, which was later superseded by the American-origin name *Sequoiadendron*), planting them in large English gardens became quite fashionable. To this day, the sequoia's characteristic bell-shaped crown and massive, reddish bark-boles can be seen near many old manor houses on great estates, as well as in botanical gardens across Europe.

A human view

Like a first glimpse of the Grand Canyon, as you closely approach sequoias, their great size becomes increasingly real. Your senses tell you it is of a scale exceeding anything previously seen, but you can't quite believe it, at first. It is only when you stand at the base of one of the largest sequoias and tilt your head far back to follow the great column upward, that you begin to realize their enormity.

Indigenous people of North America undoubtedly were the first humans to see giant sequoias. Little of their oral history is known to us, but we have a Miwok name for them, "woh-woh-nau," which is intended to approximate the sound of an owl, the guardian spirit of the tree. This name was slightly modified by white settlers to "wawona"— a name that was later given to a particular giant sequoia tree with a tunnel carved through its base, located in what is now known as the Mariposa Grove, Yosemite National Park. Countless visitors in carriages or automobiles posed for photos in the Wawona Tunnel Tree, until it toppled in 1969 because decades of traffic around the tree had damaged its roots.

The enormity of the largest sequoia trees is also difficult to express in words. Perhaps that is partly the reason that original published descriptions were mostly ignored, seeming too fantastical to be believed. The first written report was by a chronicler of the 1833 travels of a hunting party led by Joseph Reddeford Walker. Various other accounts appeared after that but received little attention, until an 1852 account by a hunter named A.T. Dowd was publicized in newspapers. In any case, after 1852 the world took note, and many visitors came to see for themselves. In addition to describing the great size of the trees, visitors began counting tree rings on stumps and reporting ages exceeding 2,000 years.

Destruction and enterprise

Within a decade of the first botanical and newspaper reports of giant sequoias, entrepreneurs decided to show people, rather than just tell them. In 1859 they "skinned" one of the largest trees in the Calaveras Grove and transported the bark-skin of the "Mother of the Forest" to London where the lower 110 ft (34 m) was reconstructed inside the grand hall of the Crystal Palace. Subsequently, cross-sections of other felled giant sequoia boles were cut and shipped to dozens of museums and science institutions all around the world.

For millions of people who had never been to California, their first experience of the enormity of giant sequoias was viewing a towering, circular slab of wood from a tree stem with a timeline of historical events traced to the tree rings. There are, however, multiple meanings of these great circular museum displays. They may represent the awesome natural history told by trees and their rings, but they remain a sad testament to the destruction of trees and forests by European colonists and settlers.

The most famous examples are the giant sequoia cross-sections in the American Museum of Natural History in New York City, and the Natural History Museum in London. Both sequoia sections are from the same tree cut down in 1891 at the behest of the American Museum. The hubris of felling a 1,400-year-old tree for a museum display was massively exceeded by that of lumbermen, who at the same time were cutting nearly one-fourth of the largest and oldest sequoias in the world for profit.

Most of the Sierra
trees die of disease, insects, fungi,
etc., but nothing hurts the big tree. I never
saw one that was sick or showed the slightest sign
of decay. Barring accidents, it seems to be immortal.
It is a curious fact that all the very old sequoias had
lost their heads by lightning strokes. 'All things come to
him who waits.' But of all living things, sequoia is
perhaps the only one able to wait long enough to make
sure of being struck by lightning ... at present only
fire and the ax threaten the existence of these
noblest of God's trees.

JOHN MUIR, *THE YOSEMITE*, 1912

A silver lining?

There is no ethical justification for killing multi-millennia-old trees found only in small, scattered groves in a single mountain range. But the existence of such magnificent trees as shown by the great exhibition and museum displays of the late 1800s resulted in a broad public awareness of them. The wholesale destruction of the trees by logging enterprises stirred the conscience of conservationists and the public. This led the United States (U.S.) government in 1890 to establish the Sequoia and Yosemite National Parks, the second and third such parks to be created in the country. The chief goal was to protect the largest remaining and most scenic giant sequoia groves from further indiscriminate logging.

Another slightly mitigating outcome of the sequoia logging era was the scientific resource made accessible in the thousands of remnant stumps left behind. Those rooted tree-ring archives have ultimately yielded a treasure trove of climate and fire history information; specimens collected from sequoia stumps were, and continue to be, central in the development of dendrochronology, providing insights about drought and wildfires over the past several millennia. Recent hot droughts and wildfires are now killing many of the remaining largest sequoias. As the climate continues to warm perhaps the stories that tree rings from old stumps are telling will help guide us in saving the remainder of living giant sequoias.

A first look at tree rings

The first scientist to study giant sequoia tree rings was Yale geography professor Ellsworth Huntington (1876–1947). He was a proponent of "environmental determinism," arguing that climatic variations explained much of human history. Inspired by Andrew Ellicott Douglass's original 1909 studies of rainfall correlations with tree rings in Arizona, Huntington traveled to California's sequoia groves in search of a long climate record from tree rings.

Over two summers in 1911 and 1912 Huntington and his assistants climbed up on cut stumps of giant sequoias. Using a pen knife, hand lens, and ruler, they counted the number of rings per decade and measured the total ring width per decade on about 450 stumps. They counted the rings of seventy-nine trees that were more than 2,000 years old, and three that were more than 3,000 years old. The oldest tree had 3,150 rings.

Dendrochronology expands (and solves a mystery)

Initially following Huntington's directions to the oldest sequoias found, Andrew Ellicott Douglass's subsequent visits and studies in the Sierra Nevada became instrumental to the understanding, recognition, and description of the guiding principles of dendrochronology. Building on Huntington's and then his own work, he was able to articulate the fundamental dendrochronology concepts of accurate dating (using crossdating), and site and tree selection to maximize climatic sensitivity.

Douglass visited the giant sequoia groves five times, collecting specimens between 1915 and 1925. One of his goals was to obtain samples to develop an accurately dated annual time series of giant sequoia ring growth that he might use to crossdate ancient timbers from the southwestern United States. Archaeologists were excavating the "Great House" pueblo ruins at Chaco Canyon, New Mexico, at that time, and they had asked Douglass to try to determine when the trees were harvested, thereby allowing them to date the pueblo construction.

From 1914, archaeologists sent Douglass timbers from dozens of different pueblo ruins. He also collected tree-ring samples from living trees across the southwestern United States. He was able to crossdate the ring-width patterns in the old timbers against each other, but not with the tree-ring patterns from living trees in the region that were up to about 500 years old. Apparently there was a "gap" between the earliest rings of the oldest living trees and the latest rings in the timbers. He hoped that the multi-millennia giant sequoia tree rings in California would bridge this gap.

However, the climate variations and hence ring-width patterns were too different between California and New Mexico. He was unable to see a match (a crossdate) of the pueblo tree rings with the 3,200-year giant sequoia tree-ring chronology that he developed from his collections.

Douglass eventually found the "missing link" between the "floating" pueblo timber chronology and the southwestern United States living tree chronology in 1929. Archaeologists excavated the missing link timber from a pueblo ruin in Show Low, Arizona, and Douglass was present at the time. That evening by kerosene lamp, he crossdated the ring patterns and bridged the gap, and for the first time the Great Houses of Chaco were dated to the calendar years of construction, circa 600 to 1130 Common Era (CE).

From Viking ships to the Curve of Knowns

Douglass's archaeological findings were published in *National Geographic Magazine* in 1929, establishing the power of the dendrochronology method, which has now been used around the world for dating everything from sunken Viking ships to Stradivarius violins, and from droughts to forest fires.

Although giant sequoia tree rings did not directly aid in solving the age mystery of the Great Houses of Chaco Canyon and other pueblo ruins of the American Southwest, they later served in a momentous breakthrough in geo-chronological dating. Willard Frank Libby (1908–1980) was awarded the Nobel Prize in Chemistry in 1960 for his discovery and demonstration of the use of carbon isotopes for determining the ages of long-dead biological materials (bones, wood, etc.) in geological and archaeological contexts.

In testing of his estimated half-life of the unstable carbon isotope 14 (C14, 5,730 +/- 40 years) he measured C14 amounts in a set of materials of known ages. These constituted his "Curve of Knowns," which showed the C14-based date estimates were relatively accurate. He used measurements of C14 in the wood from inner rings of the giant sequoia "Centennial Stump" provided by the University of Arizona Laboratory of Tree-Ring Research (LTRR)—rings dated to the tenth century BCE by Douglass—as one of the earliest and key dates in his Curve of Knowns.

How my own research began

In 1987, I was a freshly minted PhD student at LTRR with a deep interest in the ecology and history of wildfires. Prior to my graduate studies I was a wildland firefighter in the Gila Wilderness, New Mexico. My master's thesis was on the fire history of a part of the Gila Wilderness. That work involved collecting cross-section samples from so-called "fire-scarred" trees. These characteristically are trees with charred, basal wounds caused by repeated fires burning around the tree but not killing it. In the case of my Gila studies, the fire-scarred trees were primarily ponderosa pine (*Pinus ponderosa*), and the oldest were about 500 years in age.

One of the key findings from my Gila fire-scar studies, as had been found in other fire history research, was that ponderosa pine forests used to sustain frequent, low-intensity surface fires. Widespread fires recurred about every 5–15 years until about 1890–1900. Then they ceased burning as frequently or as extensively as they had for centuries before. The initial cause of the cessation in most cases was intensive and extensive livestock grazing. Sheep, cows, and horses consumed grassy fuels that previously carried the frequent fires. The seasonal "driveways" created by massive herds also disrupted the fuel continuity and hence

the spread of wildfires. After 1910, the U.S. Forest Service, National Park Service, and other agencies began putting out all wildfires as quickly as they could as a matter of policy.

In 1942, the iconic animated film *Bambi* was released by Walt Disney Studios. It featured a dramatic scene of a wildfire ignited by careless hunters that killed Bambi's parents. The massive success of the film in depicting human-set fires as evil convinced the U.S. Forest Service that they needed their own cartoon animal for their fire prevention propaganda program. Thus, Smokey Bear was introduced in 1944, with the unforgettable mantra: "Only you can prevent forest fires!" Smokey was arguably the most successful advertising gimmick in history, leading to decades of effort not only to prevent careless human-set fires, but to extinguish all wildfires regardless of origin, including those set naturally by lightning.

Reaching out to the sequoias

Like ponderosa pines, giant sequoias also form fire scars at their bases. In fact, virtually all the oldest giant sequoias have basal cavities formed by repeated fire scarring. I had learned the fact that sequoias had fire scars when reading Douglass's writings about his sequoia collections and studies that clearly showed massive basal fire scars. I dreamed of the possibility of one day collecting fire-scar samples from giant sequoias.

My chance came when I heard about a controversy involving the use of prescribed fire (manager-set fires) in the sequoia groves. This change in policy from the Smokey Bear paradigm that all fires were bad and must be extinguished came about in the 1970s and '80s as ecological evidence mounted that some kinds of fires were necessary for forests to regenerate and to prevent the accumulation of too many trees—living and dead. One of the fires set by managers in a sequoia grove to reintroduce fire as an ecological process had burned relatively hot, causing scorching of the lower stems of a few big trees, and killing some of the understory firs and pines. This led to a call for stopping the prescribed burning until more fire history research was done. Previous researchers had conducted fire-scar studies within sequoia groves, but they primarily used fire-scar samples from firs and pines, so the histories were only a few hundred years long.

I contacted fire scientists at Sequoia-Kings Canyon and Yosemite National Parks. I told them about my abiding interest in working on giant sequoia fire history, and the fact that Douglass's collections of sequoia tree rings and his files at LTRR provided a perfect basis for finding the oldest trees and stumps with fire scars, and tree-ring crossdating samples from them.

With grant funding from the National Park Service (and later the U.S. Geological Survey), we began our preliminary collections and studies in the Mountain Home Grove of giant sequoias. This grove is within a state forest and county park today, and it had been extensively logged from the late 1800s into the twentieth century. There are many stumps there, including two of the oldest trees Huntington had ring-counted.

Douglass had also collected samples at Mountain Home, and so we used his field notes and maps to find the stumps that both he and Huntington had sampled. They were usually quite distinctive because both scientists had carved their tree identification numbers onto the cut surfaces of the stumps. Also, Douglass's collections were evident as "V-cut" notches where radial sections had been laboriously cut with long cross-cut saws.

A formidable start

Like countless visitors before us, when my team first arrived we were awestruck by the magnificence of giant sequoia trees and their groves. The oldest sequoias tend to show their age with certain characteristics: gnarled bases, often with multiple fire scars; huge branches; and often with dead tops appearing as a spike rising above the uppermost green foliage. Intrepid tree-climber scientists have now ascended many of the largest and oldest sequoias. Photos show them sitting on the great dead pillars of wood at the top.

Some of these dead tops are hollow. I'll never forget our wandering and working in the groves and occasionally hearing a large pileated woodpecker hammering away on the outer shell of a dead sequoia spike top. The hammer blows of the woodpecker were magnified by the great wooden drum, echoing throughout the grove. Initially we were unsure if we could obtain samples from the sequoia stumps or logs because of

their massive size. And of course, we had no intention or permission to try to take large partial sections from living sequoias. We were quite experienced as sawyers using gas-powered chainsaws to obtain partial and full sections from pines and other conifers. But giant sequoias are a different beast. From observations of a full sequoia cross-section mounted in the parking lot near the largest tree in the world, the General Sherman tree, we could see that fire-scar cavities extended deeply into the bole. Our first thought was, "We're gonna need a bigger saw!"

Down in the "mines"

Ultimately, we spent five summers in the Sierra Nevada, collecting partial cross-sections from giant sequoia stumps and logs using very large chainsaws with bars up to 7 ft (2 m) long. We collected hundreds of slabs of wood cut from the edges of fire-scar cavities near ground level of the dead trees. Each slab was 3–4 in (7–10 cm) thick and up to 5 ft (1.5 m) wide and long. These partial sections were initially very heavy since the wood was typically near saturated with moisture. Huge wheeled containers and many volunteers helped us haul the heavy sections out of the groves. We filled large rental trucks every summer and brought the wood to our laboratory in Tucson, Arizona.

Collection of the big slabs of wood was incredibly hard and dirty work. We looked like coal miners wielding chainsaws, pry bars, and axes. Our "mines" were the deep, blackened cavities in the great tree boles. Charcoal and dirt covered our faces and clothes, with sweat leaving streaks down our foreheads and arms. Every slab of wood required an hour or more of cutting, prying, and pulling. At last, the section would break free, and we withdrew it in rising excitement to see what kind of fire-scar record was buried deep in the wood. Sometimes we could see only a few scars amid the hundreds of tree rings. And sometimes there would be twenty, thirty, forty, or more separate fire-scar events revealed as black streaks within the tree rings, exposed to view for the first time in a thousand or more years. We shouted in joy during these moments of discovery!

In addition to the large slabs cut near ground level, we also obtained what we call "radial sections." These were narrow slices taken up higher on the stump to avoid ring distortions caused by fire scars.

The radial sections were later used by my colleague Malcolm Hughes and others to develop a network of new giant sequoia tree-ring width chronologies. Along with Douglass's original collections and tree-ring measurements, these chronologies were useful in reconstructing a history of extreme drought events in the Sierra Nevada over the past 2,000 years.

Thriving from fire

Our years of field collections and laboratory work on the giant sequoia fire-scar sections resulted in fire chronologies from five different groves: Mountain Home, Atwell Mill, Giant Forest, Big Stump, and Mariposa. Later, we also collected and dated a set of fire-scar specimens from the North and South Groves at Calaveras Grove. We sampled sixteen to fifty trees in each grove, and one to five partial cross-sections were cut from each tree.

It turned out that giant sequoias are the all-time champions of fire-scar-recording trees. On the 20-ft (6-m) diameter display section in the Giant Forest we had access to the entire cross-section. On this section we dated eighty-three fire scars over the period spanning 260 BCE at the pith to 1950 CE when the tree was felled. We also discovered that giant sequoias regularly increase their growth for a period of years following each fire event. These growth surges are probably due to a combination of nutrient release in soils by the burning of accumulated leaves, branches, and grasses, and increased moisture reaching the roots. Counting both fire-scar events and probable fire-associated growth surges we counted 125 fire events recorded by this single tree over its 2,200-year lifespan.

Overall, the giant sequoia fire-scar record in the five southern groves we sampled was well replicated over the past 2,000 years. A total of ninety-seven sequoia stumps, logs, and snags (standing dead trees) had tree rings at least back to 1000 CE, forty-two had tree rings at least back to 1 CE, and five trees had tree rings back to at least 1000 BCE. Of the five trees we sampled that were more than 3,000 years old, three were previously unknown. The oldest tree we sampled was a huge snag in the Giant Forest that we called "Old 100" (after its nps identification number). This tree had an inside date of 1239 BCE and an outside date of 1844 CE The oldest fire scar on this tree dated at 1125 BCE.

Fire and drought patterns emerge

When we originally compiled the fire-scar records from the hundreds of partial sections from dozens of trees in each grove, several temporal patterns became obvious. We noticed that the highest frequency of fires occurred during the Medieval Period from about 900 to 1300 CE. Many relatively smaller extent fires occurred during that period within groves, compared to before and after. After 1300 CE, intervals tended to be longer, but fire events were more synchronized between trees within groves.

Synchrony of fire-scar events within a grove generally signifies more widespread fires. We also noticed considerable synchrony of fire events between groves: The same fire dates recurred again and again when comparing the chronologies of fire from different groves. Most of the groves we sampled are probably too distant for fires to have spread between them.

As you might expect, these highly synchronous fire dates among the groves correspond very well with the driest years. We were able to detect this close coupling between regionally synchronous fire years and drought because, in addition to the fire-scar record, we have independently reconstructed drought and temperature histories from tree-ring width chronologies. My tree-ring colleagues have painstakingly compiled tree-ring width chronologies from drought-sensitive trees in dozens of stands in the Sierra Nevada. These chronologies were calibrated with rain gauge records and temperature records to reconstruct separate histories of drought and temperature variations over the past 1,200 years.

We found highly significant correlations of these climate records with synchronous fire events and fire frequency patterns in the sequoia groves. Dry and warm conditions were associated with increased fire frequencies in groves (2- to 5-year fire-free intervals in areas of about 25–179 acres [10–70 ha]) and wet and cool temperatures were associated with reduced fire frequencies (15- to 30-year fire-free intervals).

As I mentioned before, the highest fire frequencies occurred during the Medieval Period. Historical evidence from disparate parts of the Northern Hemisphere shows it was unusually warm during this time. Several lines of evidence also point to a "Medieval Drought Period" in western North America, including low lake levels and increased fire activity in the Rocky Mountains as indicated by studies of charcoal accumulation rates in sediments. A general pattern of increasing fire frequency from around 900 CE to about 1300 CE, and a decline, thereafter, was also confirmed by sedimentary charcoal studies carried out by Scott Anderson and colleagues in the Sierra Nevada. They counted charcoal particles in sediment cores extracted from wet meadows in some of the same groves we studied with tree rings and fire scars.

Here come the humans

Although tree-ring and sedimentary studies show that annual to century-scale variations in climate explain a significant amount of the variation in fire synchrony and frequency in giant sequoia groves, there was clearly another important driver of the changes in fire occurrence in specific places and times—humans. It is well known from oral histories, and from historical and anthropological research that Indigenous people of the Sierra Nevada used fire for many purposes. And there is abundant evidence of the presence of people in varying numbers, including both seasonal inhabitation within the groves, and permanent villages nearby.

Fire-scar patterns and their changes over more than 2,000 years in giant sequoias might be used to explore the changing role of human-set fires, but these studies have not been conducted yet.

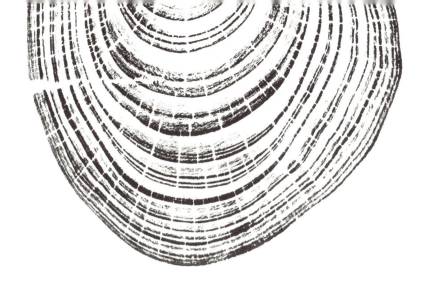

A fraught future?

The role of climate and humans, and their interactions, in controlling past fire regimes is of more than academic and historical interest. If we are to manage and preserve giant sequoias into the future—as climate continues to warm due to rising greenhouse gases in the atmosphere—our understanding of these interactions will become increasingly important.

The urgency of working to restore giant sequoia groves to ecological conditions that are more resilient to high-severity fire was shown rather shockingly during the summers of 2021 and 2022. Wildfires during these hot-dry summers swept through many giant sequoia groves, killing an estimated 13 to 19 percent of the world's sequoia population. To the best of our knowledge, this is an unprecedented simultaneous loss of ancient giant sequoias in multiple groves over the 3,000 years of tree-ring records we have studied.

For comparison, consider that loggers felled perhaps a quarter of all the large giant sequoias over a 30- or 40-year period in the late 1800s. The full impact of the recent wildfires on sequoias, and their consequences for regeneration, are yet to be fully assessed. In any case, it is evident that we are losing many multi-millennia-aged trees, and it is only a matter of time until some of the smaller, isolated groves will lose all their standing mature trees in severe wildfires.

The recent killing of large and ancient giant sequoias by very intense wildfires was due to scorching and burning all the way to the tops of their crowns and burning out the bases of old fire-scarred veterans. The extraordinary warm droughts we are experiencing are clearly a major factor in driving these severe fires. Other factors include the changes in forest structure and accumulated fuels in giant sequoia groves since low-intensity surface fires had ceased burning in the groves by around 1850.

Intensive sheep grazing in and near the groves was the initial reason for the cessation of frequent surface fires, because the sheep ate grasses that carried the flames. Later, the Smokey Bear policy of extinguishing all fires kept the groves from burning for decades. Finally, fire ecology research, and especially fire history evidence from tree rings, changed our understanding of the importance of low-severity surface fires in giant sequoia groves.

Fighting fire with research

The National Park Service has recognized the well-documented changes in fire regimes and the structure of giant sequoia groves since the 1850s, and has been working for decades to restore low-severity surface fires to the groves. However, this work has been slow, in part because of the risks and difficulties of using fire and dealing with smoke in these highly visited and beloved groves. Still, the positive effects are evident: The recent wildfires tended to become much less intense when they encountered recent prescribed burns, and less overstory mortality was experienced.

A lesson from tree rings and fire scars is that ancient giant sequoia trees and groves can withstand wildfires when they occur very frequently at low intensities over multiple decades. The frequent fires maintain fewer trees in the understory and low fuel amounts on the ground. Fires are less intense when they burn during hot droughts under such reduced fuel conditions.

Lightning ignitions are insufficient to ignite fires at the frequencies required to maintain such fire frequencies within the groves, especially now in modern landscapes. Managers will need to keep adding fires to the groves in the right times and places. Both modern science and Indigenous knowledge of how to use fire in giant sequoia groves will be needed in the coming hot decades.

Standing tall

While it's clear that a giant sequoia standing today is under serious threat—by our changing climate, and by increased human activity—its ability to survive can also not be underestimated, just as its immense stature cannot be understated. For those standing close to such a tree and taking in its enormity—its height, towering branches, and sheer girth— it is a true marvel to comprehend the lifetime of wildfires, changing seasons, and revolving centuries it has endured.

Our oldest giant sequoia today was certainly standing tall when Indigenous peoples such as the Miwok made their summer camps in the mountains, trading obsidian and berries with other tribes 3,000 years ago. The tree was 3,000 years old as the Gold Rush raged around it, much of its family toppled by colonists and settlers in their scramble for riches. Its branches had already withstood several dozens of wildfires, outlasting the Maya Civilization, the time of Elizabeth I, and the Little Ice Age that saw the Thames River in London freeze over. The tree remained, quietly growing, surviving, in one of the small groves we can still see today.

It gives hope that, long past our lifetimes, the sequoias will do just that—continue to tower high above the forest, their trunks gnarled but alive, each year building ring upon ring, offering a beacon of resilience in uncertain times.

· · ·

2

···

BOSNIAN PINE

MOMCHIL PANAYOTOV

SPECIES

Bosnian pine, *Pinus heldreichii*

············

LOCATION

Southeastern Europe—Mountains in the
Balkan Peninsula and Southern Italy

············

ESTIMATED AGE

Up to 1,100 years old

The Bosnian pine is a symbol of resilience. These magnificent
trees might not grow as old as the bristlecone pines or as large
as the redwoods in western North America, but in the
mountains of southeastern Europe they are a symbol of
a sturdy tree that can withstand the harshest of mountain
conditions. Living at the subalpine and tree line zone, often
on sun-exposed and very steep, rocky slopes, this pine has
learned to survive almost anything—cold, drought, poor soils,
ground fires, avalanches, even people starting to cut it and
then leaving it quarter-cut if they did not like the width of
the heartwood. But it is also a symbol of beauty, with its
frequently spiraled stems, covered with almost-perfect
hexagonal scales, supporting large branches often as big as
other trees and sometimes providing scarce shade to one of its
favorite companions—the delicate Edelweiss.

First found on the mountain of the Gods,
Mount Olympus in Greece, Bosnian pine is a symbol
of resilience and obstinate beauty.

Finally, after hours of climbing, I am at the top of the steep slope above the high band of white marble rocks. In this especially strenuous last hour, I have managed to perform the strange dance of jumping from one leaning dwarf pine stem to the next, without falling and sinking into the maze of branch-like stems hanging a few feet above the ground. This strange jumping-balancing-jumping dance is the only way I know of to pass through a *Pinus mugo* bushland if I want to avoid crawling over the ground, and it comes with a high risk. A fall could mean heavy injury and many hours of torture for the mountain rescue service volunteers who would have to carry me by hand through these same bushes. Will my dance be worth it? Looking from the bottom of the Bunderitsa valley in the Pirin Mountains, toward my favorite opposite slope of Todorka peak, I dearly hope so. I remain with the impression of scattered trees with large crowns that frequently have dry tops—a potential sign of old age. And my aim for this mini-expedition has been to find old trees. I search for them out of a pure desire to learn more about the majestic Bosnian pines (*Pinus heldreichii*).

A quest for something special

Yes, being a Bulgarian and a trained forester I knew the story of the ancient Baikusheva mura tree of this species, discovered a century ago during the first scientific expedition in these mountains and believed to be 1,300 years old. But after my passion for dendrochronology started to expand, and after looking at numerous trees without finding any that old, I had doubts. In addition, Stefan Mirchev, the professor who sparked my interest in dendrochronology, was quite skeptical on the dendro-chronological value of this species. "I am sure I have seen a core of one tree older than six hundred years," he said, "but if the famous Fritz Schweingruber from the Swiss Federal Institute for Forest, Snow, and Landscape Research (WSL) did not find anything special in these tree rings, then you will also not find it."

Well, I know both of these men, and trust them, but I had also seen the measurements of the cores that Fritz Schweingruber managed to collect in the 1980s. They were from young trees, fast growers, probably at the bottom of the valley close to the road. And it came as no surprise that he had such cores—these were the communist years and surely a Western researcher wandering around and doing something to the trees would be met with suspicion, which could result in hours of inter-rogation at a local police station. I was positive that Schweingruber did not have the opportunity to fully explore the forests and look for the oldest and most interesting trees.

A glimmer of hope

Back on the slope. The first glimpse of it is impressive. A small herd of wild goats enjoys the tranquility and lack of people and slowly searches for tiny patches of grass among the rocks. The Bosnian pine trees are also impressive—crooked stems, irregular crowns, some widely spaced, some clumped as if deliberately planted together. In fact, they were—not by people but by birds making stashes of seeds for the winter and then forgetting them, or by seeds finding the best place to grow on old

BOSNIAN PINE TIMELINE

• • •

Key climatic events recorded by Bosnian
pine trees on the Balkan Peninsula

941 CE First dated ring of Adonis, in the Pindus Mountains

1620s Little Ice Age—low growth of Bosnian pines

1660 Low tree growth in Bulgaria related to cold, wet weather

1724 Major fire year in the Pirin Mountains

1730s Low growth of Bosnian pines during cold period

1883 Major avalanche year recorded by trees in the Pirin Mountains

1946 Dry summer resulting in low growth of trees and numerous fires

1963 Low North Atlantic Accilation (NAO) phase causing harsh winters

rotting stumps and logs. Which way to go—to the right or to the left? I have only a few hours before I must go back and find my way to the safety of the path and the valley below. And I must be very careful because I know the slope is increasingly steep below me and a fall could result in rolling and falling from high rocks. Deep in my backpack I have two pieces of rope that together are not more than 100 ft (30 m); I know their long history and that they should not be trusted too much. Perhaps I should head for the rocky ridge to the right?

Reading books and papers about ancient bristlecone pines and other old trees, I have learned that if you are looking for old trees, the chances of finding them are always higher at rocky locations without thick soils. I head for the ridge, carefully traversing the steep slope with very thin soil and rolling rocks, soon reaching it. The first tree I choose has a wide strip of dead bark and is spiraled as if an ancient giant has twisted it two or three times before giving up this funny game. I place the borer tip between two of the almost perfectly hexagonal scales of the bark (a feature that has led to some people calling the species "turtle pine"), press the metal holder firmly, and start the slow clockwise rotation. The hardened steel winds its way deeper into the stem, smooth with every rotation. There is no sense of that hollow feeling when coring rotten conifers. I have spent so many hours trying to clean the borer after mistakes of coring half-rotten trees that I do not want to repeat it, especially today. Soon I reach the point where I can no longer rotate the borer, so I carefully insert the tiny piece of metal that is cleverly called an "extractor," make the mandatory one rotation counterclockwise, and carefully pull out the extractor with the core in it. It is the moment of truth.

The first thing I notice is the smell. The sweet, yet fresh smell of pine resin, but more citrus-like than other pines I have cored. Then I see the rings. They are so tiny and so manifold. With a trembling hand, I pull out the whole core and start a quick count. One hundred, two hundred, three, four, five hundred; yes, definitely at least five hundred tree rings, each produced during one of the summers in the past. There will be old trees on this rocky ridge today! I carefully put the core in the storage board, write some notes, and take the mandatory picture. I put my hand thankfully on the stem of the pine, thinking for a few seconds of the many hundreds of years it has endured the strong winter winds, heavy snows, pouring rains, and exhausting droughts. These trees are remarkable.

Obstinate beauty

First found on Mount Olympus in Greece, this species is a symbol of a transcendent organism surviving from a time when trees easily lived more than 1,000 years. And here it remains—growing at a slow pace, withstanding each challenge. Bosnian pines are among those organisms one would call "stubborn" if they could only be described with one word. They grow in the harshest conditions found around here, where almost no other plant wants to grow. Most are on steep rocky slopes rooted to the shining white marble. Marble is beautiful, white as newly fallen snow and often shining as if there are hundreds of miniature diamonds hidden in it. Yet this beautiful rock is not an easy one to grow on if you are a plant. Firstly, it is hard. It is old limestone that sank during great movements of the Earth core, compacting under huge pressure before emerging again to the surface. Secondly, it is built of calcium carbonate, which, when it erodes, creates a thin and poor soil. Many plants cannot survive on this soil. And to make things even worse, calcium carbonate is slowly dissolved by water, forming many tiny channels in the rock that drain water to the caves deep below. This means that the thin soils dry up soon after it rains and, if the sun is strong, which it often is in our Mediterranean climate, the plants will experience real drought. Besides all this, up here at the tree line, the summers are short and it can also be cold, especially in winter. Just a few hundred feet higher, there are no more trees—they cannot endure the harsh combination of strong winds, poor soil, and low temperatures.

"May you live many more years and please give me some of your stamina," I say as I pull my hand off the bark and I move to the next tree several feet below. It is similar. Some more trees later, I come across an old pine with a dead top and branches stretched almost like a cross.

It is neither tall nor very thick, but seems like an ancient prophet looking calmly at the valley below. A little way from this is a sheer 330-ft (100-m) drop down which I'll fall if I miss a step. I must be careful. In my notebook it becomes "tree 777." "Will you be seven hundred years old?" I wonder. I practice the same careful coring, with its familiar sweet resin smell, gently pull the extractor, and look in awe at the core. I can hardly see the rings. I give up counting after it reaches over five hundred years, admitting to myself that I see too little and there is no use in trying to count the rings. Later, the good optics of the stereoscope will help me measure the ring widths precisely on the specialized table and I will get my answer.

Difficult times maketh the tree?

My eyes already catch another tree, growing on a small terrace between the rock bands. There is only one word to describe it: majestic. It is thick, with branches so wide it's as if they are separate trees peeking out of the main stem. I want to reach it, to touch it at least. Time for the ropes. A few minutes later, I am there. I pull the borer out of my backpack, assemble it, and start the ritual again. The rotation is smooth, but becomes harder and harder and, suddenly, I feel a sharp tremble in my hands and the borer starts to rotate almost freely. Did it break? Please no, I borrowed this borer from a friend in Switzerland, since we did not have a borer of that length in our university. How would I explain a broken borer to him, and he to his bosses? I start to rotate backward, put on some pulling tension, and slowly the metal cylinder comes out to confirm the unpleasant truth. It is broken and the tip remains somewhere deep in the old pine. It has become like one of those thousands of bullets that hit the trees at the frontline of the First World War in the Alps and remain in the stems to become hidden monuments of those sad years.

Who knows, maybe I also aimed the borer at an even older bullet, a remainder of the turbulent past of these mountains centuries ago, when the local runaways, called "haiduks," hid on these steep inaccessible slopes. From the high rocks, haiduks would monitor the valley below and signal to others if there was a caravan potentially worth robbing. Those were difficult times; many men ran to the mountains to live a bandit life. They were often escaping certain death after offending some of the Ottoman rulers, who had a different religion and did not much prize the life of local Christians. In those times, mountains were a territory of harsh shepherds and haiduks, and people of the valleys looked at them with awe and fear, creating legends and singing songs of the brave free men who dared to live among the ancient trees together with the wolves and bears. Perhaps this fear of the mountains, combined with the steepness of the slopes, helped these old pines survive to our days, and to escape the logging that killed their longevity rivals—the mighty oaks of the valleys.

An obscured past

In fact, the story of the Bosnian pine is mysterious even from the moment of the first scientific interest in it. It was first described in 1863, relatively late compared to many other European species, when Herman Christ published a short article about the features of a pine that he only knew from samples received from Edmund Boissier, the author of the famous *Flora Orientalis*. Boissier himself received the materials from his friend Theodor von Heldreich, who, at that time, was the director of the botanical garden in Athens. In fact, Theodor von Heldreich had stumbled upon this pine species ten years earlier in 1851, while on a scouting mission for plants on the slopes of Mount Olympus, the mountain of the gods.

We do not know why these two great botanists, who described so many other plants from the Balkan region, did not pay more attention to the pine, but the delay later caused some scientific confusion. One year after the publication of Herman Christ's article, and unaware of

this earlier publication, the famous Austrian botanist Franz Antoine published a description of a pine found in the mountains of Bosnia and called it *Pinus leucodermis*, which means "white-barked pine," named for the light-gray color of its young bark. In this way, one pine species received two different names and, for a long time, people considered they were two separate species—one growing in the region of Greece and Bulgaria and one growing in the region of Bosnia and Herzegovina, Albania, and Montenegro.

A victory ... with more secrets to unlock

Back in the laboratory, I measure the rings of the "tree 777" core, which, indeed, turns out to be more than 750 years old. Slowly, I put together the measurements of all these old trees, compose a common chronology, and start the further analyses. To my disappointment, the trees do not show an easy-to-read story about the climate they lived in. The intuitive expectation is that if they have grown high in the mountains, where low temperatures are the primary limiting factor to tree growth, their tree rings will reveal a record of cold and warm summers, when they had low or high growth, and formed narrow and wide rings, respectively. But this expectation turns out to be untrue. In fact, to my surprise, it seemed the trees "remembered" better, and formed narrow rings in the summers of drought. But to make things more difficult, the trees also produced narrow tree rings in some years for which the climate records clearly showed cold and moist summers. This is not a situation that a tree-ring scientist likes. It seemed there were two stories, and each one needed a special key to unlock the secrets that the tree-ring record was keeping.

Luckily, thanks to the efforts of so many people in the past, we had some hidden trump cards to play. Trees store their memories in their rings—but not only in the width of those rings. In fact, tree rings are built of individual cells, each one with its own dimensions, and the cell walls themselves also differ. As summer progresses, a tree builds narrower cells, but with thicker cell walls. This is related to the amount of mass contained in the cell walls and reflected in the density of wood. Over time, researchers have learned how to measure density variations within the tree ring by taking high-resolution X-ray images. This research has shown that the density of wood produced in late summer is a good indicator of summer temperatures, particularly for conifer species growing in cold environments. One would immediately say, "Bingo, easy story, let's all measure density to study our temperature history," but only a few laboratories in the world had the equipment to reliably measure density.

Fortunately, I had just met Valerie Trouet, a Belgian scientist who, at the time, was working at the Dendrolab at the WSL in Switzerland, one of the few laboratories with this equipment. She and her colleagues had already started doing research on the same species, collecting trees in the mountains of Albania. By joining efforts with other researchers, we later built one of the longest and most reliable maximum latewood density chronologies in Europe based on the Bosnian pine tree rings. In our search for old trees, we also found numerous well-preserved dead pines—logs that had been lying on the rocks of the Pindus Mountains in Greece for centuries. By linking the living and dead tree material, we were able to extend the living tree chronology by centuries, back to 1200 CE. One of the most curious findings during our expeditions was an extremely old, lonely tree, which was later called "Adonis" after the ancient Greek god of beauty. Paul Krusic had the chance to measure the tree rings and see that the tree was one of the oldest living Bosnian pines. He then returned with a special, longer borer, collected another core, and set the record of proven age at almost 1,100 years. Krusic had found and dated the oldest-known living tree in Europe.

A climate puzzle solved

The wood cells of Adonis and its fellow ancient Bosnian pines held many secrets, some of them with far-reaching climatic consequences. Valerie and I received a first hint of these secrets when we first met in Switzerland. When looking at the ring sequences of various samples from Bulgaria, we noticed that the ring of 1976 was often very narrow and light, and not dense. I remarked that this made sense to me, because the summer of 1976 was known in Bulgaria as one of the coldest summers in recent decades. This astounded Valerie. In her home country, Belgium, 1976 was well known as one of the hottest and driest summers on record. Until the heatwaves of 2018, 2022, and consequent years, the summer of 1976 had been the heatwave of reference in Belgium, the one to which all other summers were compared. Our Bosnian pine record, however, showed us that this same summer had been one of the coldest ones in 850 years in Bulgaria.

   We had stumbled upon a climate dipole, with consistently contrasting weather conditions between two regions, that has dominated European summer weather for what turns out to be centuries. Whenever northwestern Europe, including Belgium, suffered a summer heatwave, southeastern Europe, including Bulgaria, would be anomalously cold and rainy, like in the summer of 1976. And vice versa, as we experienced in the summer of 2024—although it was the hottest on record in Europe overall, at a whopping 34.8°F (1.56°C) above the 1991–2020 average, it was southeastern Europe first and foremost that experienced record high temperatures, whereas summer temperatures in northwestern Europe were slightly cooler than average.

As it turns out, it is the jet stream that produces this temperature dipole; more specifically, the latitude at which the Northern Hemisphere polar jet stream is positioned over central Europe in summer. The jet stream describes a narrow band of strong westerly winds that flow in both hemispheres at about 6 miles (9.7 km) above Earth's surface, the altitude of airplane flights. The polar jet stream is typically located at about 50°–60° latitude, but it often shifts north and south from its average position. When the polar jet stream over central Europe flows more northerly than average, you get heatwaves and drought in the Balkans and mild summers in the British Isles and Belgium, as was the case in 2024. When it is positioned more southerly than average, the opposite happens, as was the case in the infamous summer of 1976.

We hypothesized that if changes in jet-stream latitude create a summer temperature contrast between southeastern and northwestern Europe, and if centuries-long tree-ring density chronologies from these two regions record regional summer temperatures, then we could reconstruct jet-stream latitude changes over the past centuries by combining tree-ring records from the two contrasting regions. To be honest, it was a wild hypothesis: linking subcellular measurements of wood cell-wall thickness to hemispheric-scale winds that flow 6 miles above Earth's surface. But we tried and it worked! We combined our Bosnian pine latewood density record from the Balkans with a density record from Scotland in the other pole of the summer temperature dipole, and with a record from the Austrian Alps. All three chronologies were more than seven hundred years long and we found that, combined, they explained almost 40 percent of the variance in summer jet-stream latitude over central Europe. Eureka!

When we then looked at heatwaves in central England since 1659, when the British first started measuring temperatures with thermometers, thus creating the world's longest continuous instrumental temperature record, we found that they consistently occurred when our reconstructed summer jet stream was positioned in the south. On the other end of the dipole, we found that heatwaves in Italy and Greece, derived from early instrumental measurements, as well as historical documents, happened predominantly when the jet stream was in a northerly position. These findings and similar ones for regional rainfall and floods confirm that the jet-stream-driven dipole has been controlling European summer weather since the Middle Ages.

Expanding from heatwaves

This century-long match between our jet-stream reconstruction and historically documented summer climate extremes gave us the idea to look at other extreme events, such as wildfires and grape harvests in the Balkans over past centuries. Lo and behold, we found that when the jet stream was further north than average, and the Balkan summers were typically hot and dry, this was when most wildfires happened. In the relatively wet and cold summers of a southern jet-stream position, documents dating back to the sixteenth century recorded grape harvest delays, low grape yields, and poor wine quality. As a *pièce de résistance*, and to our great surprise, we found that the summer jet-stream latitude even played a role in the spread of epidemics and in human mortality patterns. The plague, for instance, occurred predominantly in cool and wet summers, which were associated with southern jet-stream positions in the Balkans, but with northern positions in Ireland. As a case in point, the Black Death raged in Ireland from 1348 to 1350, when our reconstruction shows that the jet stream was in an extreme, far-north position over Europe.

Our story keepers, the old Bosnian pine trees, held many other secrets of the past. If you live for so long, then you see many things happening around you. And if you live for centuries, then sooner or later a fire will occur. Lightning, an escaped spark from a campfire of shepherds, a deliberate act to chase rebelling people or clean the field of bushes so there will be fresh grasses for the sheep next summer—there could be many reasons, but once there is a flame and there are dry grasses and pine needles, the fire grows. Some of the fires become fierce and wild, reaching up to burn the crowns of trees, killing most of them. But some burn mostly the bushes and small trees, the dead branches and fallen stems, and only scorch the bark of old standing trees. Some of these trees do not survive, especially when they have thin bark, but others that have evolved to have thick bark with big scales can survive such groundfires. Thankfully, our main character, the Bosnian pine, along with its old rival and friend *Pinus peuce*, also a magnificent local pine, are among those species with thick bark.

Fire survivors—and thrivers

Wandering around the old-growth forests, I would always look at the scars on the old trees left by groundfires, counting how many were visible, and ask myself if I could try to learn more of this story. I knew of these studies of the old-growth ponderosa pines or giant sequoias in North America and how researchers used old stumps and cut wedges in the trunks of trees to read the tree rings and compose their fire histories. But this was out of the question in this forest: There were strict protections in the Pirin National Park. Even to core trees I needed high-ranking permission, and cutting a wedge out of a living tree with a chainsaw was completely unthinkable. Stumps with fire scars were too few and too rotten to use. Luckily, Dominik Kulakowski, another friend and forest researcher in the western United States, taught me a trick to solve this conundrum. By carefully aiming a borer at each individual scar and taking multiple cores of one tree, this helped to date the years of the scars and thus the fires. Not an easy task and it does not work well if there are too many scars too close to each other on one tree but, as it turned out, Dominik's trick worked quite well on our pines, which usually had only a few scars in the past four to six hundred years.

Coring for fire scars is slow work, requiring a lot of effort in ring measuring and dating to give only a few fire years as a result, but little by little our efforts unraveled a story. These Bosnian pine forests had been burning. Well, yes, no surprise there; we had some old notes from the first educated foresters, who visited these mountains and wrote that they witnessed a fire that burned for days and weeks on end and nobody could do anything. We also had the story of Konstantin Baikushev, who, finding his old favorite tree, wrote that it "obviously miraculously escaped an old fire that killed many of the other trees around." Last but not least, we now also had the years and positions of surviving fire-scarred trees and could build a story. This story told us that there were years when fires burned in different valleys in the mountains, suggesting that this was not a single small fire started by lightning or something else and affecting only one location, but that climate contributed to the fires.

It further seemed that the fires were both Bosnian pine killers and helpers—they killed younger trees with thin bark, but they also killed larger, thin-barked trees of species that grew faster than Bosnian pine, such as Norway spruce, silver fir, and Scots pine. On soils that were not poor, these species grew faster and taller than Bosnian pine, gradually overshadowing it. For a species such as Bosnian pine, which evolved to grow slow in harsh conditions, but not tolerate shadow, this was not good. Yet a fire cleaned the ground of competition, setting things to a new start, allowing the seeds of Bosnian pine to use the light to start their growth, with the logic that in the following decades the more abundant light would be more beneficial for them than for other species. It is a long story of forest dynamics, but in the nature of plant life there is no vacuum and every niche is taken.

Avalanche understanding

There were not only fires in the turbulent history of those forests. In the steep Pirin Mountains with abundant snow, there are avalanches, which can be incredibly destructive, leaving long and wide stripes of crushed trees or only soft, leaning bushes. Many forests in the Pirin Mountains had such lines along the slopes as remnants of past winters with thick snowpacks and large avalanches, such as those in the 1950s, 1960s, and 1970s. Local people feared the "white beast" and avoided going up in the mountains in winter time, but after the beginning of the twentieth century and the rise of forest exploitation, they used snow for easier transport of logged trees by dragging them with strong oxen. This

inevitably caused the collision of weak human bodies with tons of snow, broken trees running down the slopes at huge speed, and often tragic news for families in the villages below. Later came tourism—ski lifts, hikers, skiers, and snowboarders, all desiring to venture deeper into the beautiful mountains.

Accidents happened and still happen, requiring response measures, which in turn need knowledge. But how do we get to know where and how often these large avalanches happened? Without written records, such as those kept by generations of local monks in some Alpine valleys in France, and Switzerland, where could we look for this knowledge? The tree rings again. It turned out that our stubborn Bosnian pines with their flexible young stems and solid massive old trunks could also keep a story of when they were hit by an avalanche. This would either leave a scar, break the stem (but the tree would survive and regrow), or bend it or almost uproot it, thus forcing the stems to grow horizontally and produce so-called "reaction wood," all of which were clues a dendrochronologist could use to date when the avalanche occurred. Hundreds of trees, thousands of cores, and many hours of work resulted in a map of the areas of a slope hit by massive avalanches over the past centuries. This map was then used to calculate the power of avalanches if they managed to reach a certain location and get a better knowledge of the real risk.

A glorious return

Back to my dance in the forest. Sweating, a bit bruised, neck full of nee-dles, I emerge from the *Pinus mugo* bushland. One more step and I am at the top of the slope. The morning sun kisses me, and I look at some of my favorite trees. Every year I repeat the ritual—after all, I have given myself a reason. While jumping from stem to stem through the bushes, I curse myself for the decision I made years ago to install a small data logger—which records temperatures every hour—to the hexagonal scales of an old tree. But the view at the end is rewarding, and, of course, the annual data needs to be downloaded.

I take a look around. There is the old dead tree I know so well, lying still on the ground on the steep slope below. It had pink wood and it was really hard to extract the core, but provided the record for the old-est years. I see the familiar, small curled trees on the rocks at the left. I once borrowed a 260-ft (80-m)-long rope and had to hang on it to extract their cores. They were some of the oldest trees I had the chance to touch. I see the trees on the ridge to the right, where I first went. I need to walk a bit to reach them, but I often do it. There they are: "the twisted one," "tree 777," "the borer eater," and the others. I am happy to see that all are in good health. Sometimes I notice a big branch is broken by the weight of the winter snow, but at least these old ones are still alive. And luckily for them, the chainsaw is forbidden here, so even if a crazy logger was willing—or able—to come to this almost vertical slope, they would not be allowed. All in the hands of nature!

As I look out at these swathes of majestic, stubborn Bosnian pines, I call to them: "Let you live many more centuries, old friends," and start my careful way back to the path.

• • •

3

. . .

CLANWILLIAM CEDAR

EDMUND FEBRUARY

SPECIES

Clanwilliam cedar, *Widdringtonia cedarbergensis*

. . . .

LOCATION

Cederberg Wilderness Area, South Africa

. . . .

ESTIMATED AGE

400 to 600 years old

The charismatic and endangered Clanwilliam cedar has provided the longest well-dated tree-ring chronology for Africa south of the Sahara, attracting researchers from all over the world, all searching for the elusive temperature and rainfall record going back through time. This iconic tree is endemic to the Cederberg Wilderness Area, a rugged mountain reserve approximately 125 miles (200 km) northeast of Cape Town in South Africa. I first saw these trees in 1971 as a sixteen-year-old on a rock-climbing trip organized by the Cape Province Mountain Club. This trip not only cemented my lifelong association with mountains, but also with the local Clanwilliam cedar.

Clanwilliam cedars were once a dominant feature of the fynbos vegetation of the Cederberg mountains in South Africa. Today, these charismatic trees are under threat.

D ropped off at the trailhead late that evening, a group of us camped under a line of oak trees that led up to a ramshackle farmhouse on the far side of a little stream. I can still feel my excitement at seeing for the first time the reddish-orange rock walls reflecting the early morning sun of our destination—the Tafelberg—looming high above us. The cedars grow in a tight altitudinal band from just above where I was standing to just below the Tafelberg. For different reasons, both the Tafelberg and the cedars would become important to me over the next fifty years.

The rock-climbing group and I depart from our camping spot and begin our hike. The trail from the oaks takes us along the edge of fields of maize, pumpkins, and beans before we get to the farmhouse—the goats have already been let out, but we do stop to look at the piglets. This trail, connecting the farm Welbedacht with the little hamlet of Langk-loof on the northeast of the Tafelberg, was originally built to allow for donkeys to carry goods for trade over the mountains. Having started our walk at 2,625 ft (800 m) above sea level, we climb past the farm dam to our right before the trail dips down to cross a small stream. At the top of the low rise, I stop to look toward where we are heading.

Unique biome of outstanding beauty

The rocks surrounding us form part of the Cape supergroup rocks laid down as sediment between 350 and 500 MYA in several layers or formations, each defining a geological age. The rocky area we are walking through forms part of the Peninsula formation and, just before a little plateau at about 3,900 ft (1,200 m), there is a subtle change to the Pakhuis formation. It is the sandstone of the Peninsula and Pakhuis formations that are responsible for the fantastical rock configurations that make the Cederberg famous. The reddish-gray rock of these two formations may form almost any shape and size that a twisted mind could conceive of. Large boulders balance on each other, separated by cracks, grooves, and hollows of all sizes. Weathering has resulted in the surface of these rocks to be covered in knobs, pockets, and spikes. The result is an astoundingly beautiful, sparsely vegetated landscape.

The path winds up the gulley through these fantastical rock arrangements before plateauing slightly and then rising less steeply on the shale or mudstone of the Cederberg formation at 4,600 ft (1,400 m). Topping all of this is the reason why we have come here: a 650-ft (200-m) high band of bullet-hard quartsitic sandstone, the Nardouw formation, which has some of the finest rock climbing to be found anywhere.

To my left is a small rock outcrop, one of thousands, on which sits a 10-ft (3-m) high waboom tree (*Protea nitida*); looking down on the banks of the stream I can see waving restio (*Cannomois virgata*) fronds. Along with the Ericaceae, these two families—Proteaceae and Restionaceae—dominate this region. Known as Fynbos, this region is confined to an area of South Africa with a distinct cool, wet season from April to September and a hot, dry season from October to March. These two biomes in the Western Cape province, the Fynbos, and the Succulent Karoo, which together are known as the Cape Floral Kingdom, occupy just 6 percent of South Africa, yet contain almost 10,000 plant species and seven endemic families. The sandstone rocks we are walking through result in some of the most nutrient-poor soils found anywhere in the world, and in the hot, dry season, the regular fires endemic to

CLANWILLIAM CEDAR TIMELINE

• • •

The original chronology found only a weak relationship
with climate and several subsequent attempts have not
been successful at demonstrating any relationship
between the chronology and climate

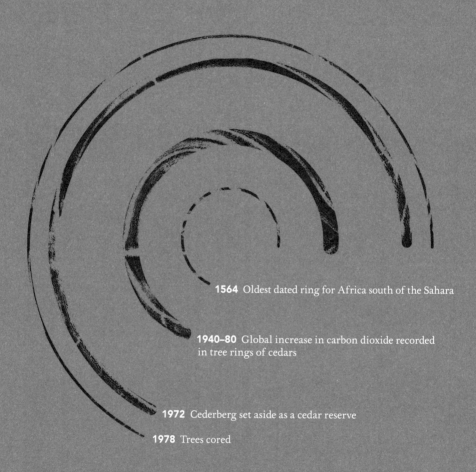

1564 Oldest dated ring for Africa south of the Sahara

1940–80 Global increase in carbon dioxide recorded
in tree rings of cedars

1972 Cederberg set aside as a cedar reserve

1978 Trees cored

fynbos keep out all trees except cedars, which are perched safely above the fires among the rocks. Several studies have attributed the high species diversity of fynbos to both the nutrient-poor soils and high fire frequency. Fynbos is said to have the finest roots in the world, allowing these species to take up nutrients where no other species may survive.

A first sighting of the tree

Dropping down to cross the stream, we climb the steep zigzag path and, at 2,950 ft (900 m), we find the cut stump of a Clanwilliam cedar—the tree responsible for the name of the mountains we are walking in. As the only largish tree in the Fynbos, cedars were exploited by early settlers, so it is no surprise to find a cut stump this close to the farmhouse. Happily, it is not long after, at about 3,280 ft (1,000 m) in elevation, that I have my first sighting of a living cedar. It is very exciting, since I had heard from my father about these charismatic trees—to then see them perched in impossibly steep locations, some standing stark against the skyline, some disappearing against the rocks, I am truly in awe. We continue our slow progress up the mountain with lots of trees on both sides of the path—many are almost growing in the stream that we cross on several occasions, but the majority perch on ledges or grow in cracks among the rocks. We drop our overnight gear at Welbedacht cave at the top of the Pakhuis formation and continue with climbing equipment up the relatively featureless, shale band to enjoy some brilliant climbing that would draw me back for several decades. I notice there are no trees growing on the shale band—the cedars seem confined to an altitude of 2,950–3,280 ft (900–1,200 m) on the Peninsula and Pakhuis formations.

To this day, I have a vivid memory of standing in the cave mouth, admiring several trees to the north of the cave. Back then, these just seemed like pretty trees. I had no knowledge of the difference between angiosperms (flowering trees) and gymnosperms (conifers, cone-bearing trees) and a very limited understanding about dendrochronology. I also had no idea of the incredible journey I would take with these cedars over the next fifty years.

South Africa's tree-ring legacy

Inspired by the work of A.E. Douglass in the southwestern United States on Ponderosa pine, a coniferous gymnosperm, several researchers in South Africa looked toward conifers to develop climate records from tree rings. There are, however, no *Pinus* species indigenous to South Africa and only seven gymnosperm species in two genera, *Widdringtonia* and *Podocarpus*. This focus resulted in the first South African tree-ring study by Hubbard in 1936 on age determination through ring counts for plantation and naturally grown Clanwilliam cedar. There was no further tree-ring work in South Africa until 1976, when the archaeologist Hall also did a ring count on a large *Podocarpus falcatus* disc housed at the Natal Museum in Pietermaritzburg. Lilly in 1977 and then Curtis in 1978 followed this up, working on the same species and concluding that age determination on *P. falcatus* was not possible because of abundant false, converging, and missing rings.

In the late 1970s, Fred Kruger, Director of the South African Forest Research Institute (SAFRI), was contacted by Val LaMarche and Peter Dunwiddie of the LTRR at the University of Arizona asking if he had any ideas for trees in South Africa with dendrochronological potential. Kruger knew that Curtis had not been successful with *P. falcatus*, but he also had a very keen interest in *W. cederbergensis*. When LaMarche arrived in South Africa from Arizona with Dunwiddie in 1978, Kruger first got them to sample some *Podocarpus*, between the towns of George and Knysna in the southern Cape, but he also took them to the Cederberg to sample *W. cederbergensis*.

The focus turns to cedars

After sampling both cores and discs at several Cederberg locations, they parked under the oak trees at Welbedacht and followed the same trail toward Langkloof I had taken seven years before. They did not stop at Welbedacht cave but continued to follow the trail up and over the little col between Tafelberg and Langberg, before dropping down to the huge old cedars at De Bos. These are some of the largest trees in the Cederberg, most likely because they are protected from the regular fires inherent to fynbos thanks to the rocks they are growing on, and the cliff behind. The location is certainly spectacular, since the little hamlet of Langkloof, which sits at the bottom of the ravine, is not visible and one can see for several miles into the Karoo in the distance. The trees stand out stark against the gray, brown rocks they are growing on and, having walked past several lesser specimens, the sheer size of these beautiful giants is impressive. Walking under the canopy, one is reminded of the spreading boughs of a massive pedunculate oak. What is particularly striking and unusual for the region is that directly under the cedars, there is no vegetation because fynbos species are adapted to high light and the cedars shade out any potential growth. I visited De Bos myself in 2001 and was lucky to see the original trees sampled by Dunwiddie and LaMarche before an unusually hot fire in 2005 swept through the area, killing several specimens.

The De Bos site yielded South Africa's first well-dated tree-ring chronology, published by Dunwiddie and LaMarche in *Nature* in 1980. The 413-year chronology, from 1564 to 1977, was based on measurements of 52 radii on 32 trees. While lobate growth and missing rings are uncommon in the Clanwilliam cedar, and circuit uniformity is generally good, there were several features that made crossdating different to any species in the United States. For example, there were many false rings, the structure of which resembled the actual terminal ring boundaries, which made the latter often difficult to determine. In addition, resin-filled bands of cells could be confused with a terminal ring boundary, and signature rings—rings that are relatively narrow common to all trees at the site—were not common. Rather, at De Bos, unusually wide rings that, in some cases, showed frost damage were more consistent between trees and could be used as signature rings. While Dunwiddie and LaMarche were able to develop a chronology, they were not able to establish a correlation between their ring-width chronology and either temperature or rainfall. Instead, they could only tentatively interpret their chronology as a record of early dry season moisture availability.

A passion for cedars grows

My first trip to the Cederberg as a sixteen-year-old ignited a passion for rock climbing that was to bring me back to the region for several decades. While there, I also developed a broad interest in the region's ecology. And I could not help but notice the decline in the number of cedars over a very short period. I asked myself if it may not be climate change and, more specifically, a decrease in rainfall amount that was resulting in the decline. I can still see myself plodding up that Welbedacht path with a heavy weekend pack on my back, asking myself why the trees lower down on the path were dying and those higher up still looked healthy. Although I tried discussing this with my fellow climbers, they really were not interested. All they cared about was that cedars made great firewood for a barbecue and the authorities had banned all fires.

By the late 1980s I had completed a master's degree in archaeology, which didn't help my understanding of cedar ecology. Yet my passion for the Cederberg and for trying to understand cedar ecology led me to a PhD in botany at the University of Cape Town. I really felt that an increase in drought frequency was the main reason for the decline in the trees, and this is when Dunwiddie and LaMarche's work directed me toward dendrochronology. A grant from the South African Water Research Commission enabled me to visit the LTRR at the University of Arizona in 1997 to learn the basics of the science. I arrived in Arizona with the cedar cores and discs I had collected from two Cederberg locations: Krakadouw in the north and Algeria in the south.

Unraveling a mystery

Under the guidance of Rex Adams at the LTRR, I developed two short chronologies from 1898 to 1994 at Krakadouw, and from 1919 to 1994 at Algeria. The trees used in this study were planted for forestry between 1897 and 1907, allowing for a good estimate of the potential age of the trees. The results of my study showed that age determination from ring counts is indeed accurate, but, in agreement with the results of Dunwiddie and LaMarche, also show no correlation between ring width and rainfall. There are several cedar discs from the De Bos chronology housed at the University of Arizona. While I was visiting Arizona, I prepared and counted the rings on one of these discs at Pakhuis Pass that gave an age of more than 600 years. This would suggest that the 413-year De Bos chronology could potentially be pushed back even further with future research on some very old trees.

While at the LTRR, I learned the techniques to extract cellulose, the carbohydrate in the cell wall, from wood in order to determine the ratios of two key stable carbon isotopes (^{12}C and ^{13}C) in tree rings. The primary reason for doing this was to determine changes in environment or atmospheric carbon dioxide (CO_2) through time. During photosynthesis, trees take up carbon dioxide through their leaves and, using water and the energy of the sun, convert this into sugars that feed the tree. As a by-product of photosynthesis, oxygen is released, emphasizing the importance of trees in supplying us with fresh air. All trees regulate carbon dioxide uptake through small openings in the leaves, the stomata. With plenty of available water, the stomata are opened, allowing for more carbon dioxide to be assimilated—as rainfall decreases, the stomata close and less carbon dioxide is assimilated. As a tree transpires freely with plenty of available water, and carbon dioxide is assimilated into the leaf through the stomata, the heavier ^{13}C isotope is discriminated against in favor of the more mobile, lighter ^{12}C isotope. Along a rainfall gradient, several studies have now shown that leaf ^{13}C/^{12}C ratios become more positive as rainfall decreases, the stomata are closed, and more of the heavier isotope is assimilated. In high rainfall areas such as the Amazon, the stomata of trees are wide open, more carbon dioxide is taken up, and, as a result, more oxygen is released, emphasizing the importance of this region to the globe.

Using carbon isotope ratios, Schubert and Zimmerman were able to reconstruct annual rainfall in Hawaii from the rings of the māmane tree. Several other studies, primarily in the Northern Hemisphere, have now shown significant correlations between ^{13}C/^{12}C ratios and amounts of rainfall. These ratios, however, may also be affected by the ratios of the two carbon isotopes (^{13}C and ^{12}C) in the atmosphere. Fossil fuel burning and changes in land use and vegetation cover have resulted in changes to the isotope ratios of tree rings laid down today relative to rings that were laid down at the start of the Industrial Revolution.

With the help of Adams, I split longitudinally eleven cores from six of the original cedars used in the De Bos chronology and pooled year by year the wood from these six trees to develop the first annual record of carbon isotope ratios for the Southern Hemisphere. I was hoping to show, as demonstrated by Schubert and Zimmerman in Hawaii, the relationship between tree-ring isotope ratios and annual rainfall. Unfortunately, I found no relationship, suggesting that water is not the main limitation for growth in Clanwilliam cedar. Instead, the De Bos tree-ring chronology I found was similar to that found in the Northern Hemisphere tracking increases in atmospheric carbon dioxide related to fossil fuel burning. We were the first to find this effect in tree-ring isotopes from the Southern Hemisphere.

Recently, an international group of scientists led by Tom De Mil from the University of Liège in Belgium measured another tree-ring parameter, tree-ring density, on samples from the De Bos site. Using X-ray computed microtomography, Tom measured the cell-wall thickness in the latewood of each ring, also referred to as the maximum latewood density, as well as in the earlywood, the minimum density. The results from this study are, however, the same as that achieved by Dunwiddie and LaMarche in that there is a weak correlation with rainfall in the early part of the dry season from October to December.

A regular dry season water supply

While there is little doubt as to the annual nature of the tree rings in the De Bos chronology, the lack of correspondence with rainfall intrigued me enough to find funding to examine this question more closely. In 1999, I was back in the LTRR, this time to learn the techniques for the extraction and processing of the water in trees. The hydrogen and oxygen isotope ratios of water extracted from the stem of a tree can be compared with the hydrogen and oxygen isotope ratios of potential source water to determine where the tree is getting its water from. What had grabbed my attention was the use of this technique in an article in the journal *Nature* with the intriguing title "Streamside trees that do not use stream water."

I chose six cedar trees at random, close to a little stream near the same cave at Welbedacht from which I had admired the cedars almost thirty years before. I instrumented each tree with a dendrometer band that allowed me to accurately measure changes in trunk diameter each month. I also installed an automatic rain gauge close to the trees so that I could compare my manual readings of the vernier at the end of each month with rainfall for that month. I also took rainwater, twig, and stream water samples for analysis of the water isotopes (H and O) at the end of each month. In this study I found that there is very little soil where the trees are growing into cracks in the rock, yet neither water isotopes nor tree growth correlate to the amount of rainfall. This study showed that the trees are not dependent on rain, but are dependent on reliable access to water from pockets between the bedding planes of the rocks they are growing on. This regular supply of water would explain how the trees are able to restrict growth to the hot dry season and would also explain the lack of correlation between the ring widths of the De Bos chronology and rainfall. During our study, South Africa's Western Cape province was hit by a severe drought (2000–2001), during which Cape Town came close to running out of water. At our study site, the little stream dried out completely and several cedars in the vicinity died. We speculated at the time that those trees that died no longer had reliable access to water and that with climate change more trees would be threatened.

A history of humans and Clanwilliam cedars

It was during this study that I realized that all of the trees I had admired around the cave mouth when I was sixteen were no longer there either. Perceived wisdom at the time was that eighteenth- and nineteenth-century over-exploitation of the tree as a timber source reduced and fragmented populations to the extent that the trees were no longer regenerating naturally. But what I was seeing was a dwindling of the number of these long-lived trees over the twenty-five years that I had been visiting the area. I struggled to see how this could be related to overexploitation more than a century earlier.

The other tree of significance in the area is the waboom, a stunted gnarled tree that grows to a height of around 16 ft (5 m). The wood of the waboom produces some timber, but it is the Clanwilliam cedar with a straight trunk of between 23 and 49 ft (7 and 15 m) that, for a long time, was the region's most valuable source of timber. The chronology at De Bos goes back to 1564, more than a hundred years before the first

European settlement at Cape Town in 1652, which would have marked the first threat to the existence of the cedar trees. The first people in the region were the Bushmen, who created some of the most remarkable rock art in South Africa dating to between 8,000 and 100 years ago. These people were gatherer-hunters, subsisting primarily from foraging and they would have had little impact on the trees. The Khoe-speaking people, known as Khoekhoe, arrived in the area around 2,000 years ago, initially with large herds of sheep and some cattle. At the start of our chronology, however, the Khoekhoe would have had large herds of both sheep and cattle, access to which was the main rationale for European settlement in 1652. The Khoekhoe moved their herds seasonally between the inland Karoo and the coast through the Cederberg. These herders would have burned the vegetation to promote new growth for their herds, in what very likely would have been the first human impact on the trees.

EDMUND FEBRUARY

The occasional fire lit by the Khoekhoe would have had little effect on the population of cedars relative to the expansion of European settlers into the region in the 1700s. The Khoekhoe social structure was destroyed by colonial settler expansion into the area, with the remaining Khoekhoe people either enslaved or incorporated into settler society. Mixed ethnic descendants of the Khoekhoe made a living by harvesting the cedar trees during the late 1700s and early 1800s, testimony of this being the pews in the church in Clanwilliam, the floors and ceiling of the Hantam Huis in Calvinia, and the farmhouses in the Biedouw valley. This early exploitation of the wood probably resulted in most of the larger, more accessible trees disappearing from the landscape. I was, however, fortunate to see the De Bos trees in 2001 and even though earlier fires had left several skeletons in the area, we were still able to walk under the enormous canopies of true giants.

It is not all about water supply

When I first walked up to Welbedacht in the 1970s, I noted that several cut stumps gave some evidence for trees at around 2,950 ft (900 m) in elevation, but that there were also plenty of trees from about 3,280 ft (1,000 m) to just below the Cederberg formation shale band at 4,590 ft (1,400 m). Over time, I noticed only the skeletons of trees lower down on the walk and the only live specimens growing at between 3,600 and 4,590 ft (1,100 and 1,400 m). The cedar growing season corresponds to the hottest, driest time of the year, from October to March. My research has shown that the cedars are reliant on a regular water supply that is made possible during the dry season through pockets in the bedding planes where water may collect and into which the trees are rooted. This regular supply of water allows for growth during the hot dry season and would also explain why no study has shown a correlation between the ring widths of the De Bos chronology and rainfall. While cedars are susceptible to drought, the enormous decline in numbers that I could not help but notice over a period of about thirty years cannot be attributed to drought alone.

Some of the most important developmental work on the cedars was done in the 1970s by SAFRI led by Fred Kruger, who directed Dunwiddie and LaMarche to the De Bos site. SAFRI set up several 1-hectare (10,000 sq m) plots to monitor the status of the cedar population. All individuals in these plots were labeled and their heights and diameters were recorded. The plots were revisited in 1977, 1979, and 1984 when the trees were re-measured, and a note made on recruitment, mortality, and probable cause of death. We revisited several of these plots in 2003 and 2005 and found a rapid decline in the number of trees in the twenty years that had passed since the SAFRI surveillance. An analysis of the fire record for the Cederberg in this time shows that the probable cause is an increase in the number and intensity of the fires to the extent that trees in rocky refugia that had been protected from fire for centuries, were no longer protected.

Some really bad news

In 2013, Joe White performed a repeat photography study in which he relocated a series of photos taken between 1931 and 1987 and repeated them. These recent photos show an equally remarkable decline in cedars over the eighty-year study period. Of a total of 1,313 living trees recorded in the historical photos across the entire Cederberg, only 387 were still living in 2013, with only 44 new recruits. Model simulations identified an increase in fire frequency and possibly higher temperatures as the primary cause of the recent high Clanwilliam cedar mortality. Trees that are still alive are now clustered in a tight band just below the shale band of the Cederberg formation at 4,590 ft (1,400 m), with most trees below 3,280 ft (1,000 m) having died off.

The Cederberg mountains were declared a wilderness area in 1973, only a couple of years after I first visited them. This wilderness declaration gives the area the highest conservation status in South Africa. Despite this, two major fires in 1985 and 2013 swept through the entire reserve and killed hundreds of trees. The cedar population continues to decline despite all management interventions.

CLANWILLIAM CEDAR

Anthropogenic warming over the past fifty years cannot be ignored as a major contributor to a decline that has resulted in Clanwilliam cedar being listed as critically endangered in the IUCN Red List of Plants. This warming has resulted in more frequent, hotter fires and more frequent droughts that, combined, negatively impact the Clanwilliam cedar's survival.

The De Bos chronology is the longest tree-ring record for Africa south of the Sahara and, since the wood decays so slowly, there is also the potential to use dead wood to push this chronology back even further. This tree is under threat in the wild and it is only management intervention that can save it from becoming extinct in its native habitat. A long climate record for the region is desperately needed for us to understand and possibly mitigate the human impact on these extremely charismatic trees. Circumstantial evidence would suggest that it is increasing temperatures and drought frequency affecting the trees. We really need to confirm this so that rewilding programs are able to focus on areas that are higher and wetter than where the trees are now growing.

Growing hope for change

Undeterred by forty years of negative results, a group of French researchers led by Jonathan Barichivich have, very recently, had me walking the Welbedacht trail one more time. This time, it is to core several Clanwilliam cedars, with the prime objective of updating the De Bos chronology from 1976 to 2024, and to use tree-ring stable isotopes to understand the physiological response of the cedars to increasing temperatures, drought, and carbon dioxide. I am pleased to know that research for the elusive climate record from the De Bos chronology—with its focus on this incredible cedar—is ongoing, bringing together dendrological researchers from around the globe. I really hope that someone is successful soon. Others are also helping to preserve the Clanwilliam cedar. Together with civil society and local farmers, a government agency known as Cape Nature—Cederberg

reserve's custodians—has insti-
gated a rewilding program that
could help sustain the current
population of trees.

This tree is under
threat in the wild and it
is only management
intervention that can save
it from becoming extinct
in its native habitat.

It is now 2025, more than
fifty years after I first walked up
the Welbedacht path and became
entranced by this charismatic tree.
The Welbedacht farmhouse is no longer
there and, unless one looks for it, the dam,
too, is gone. The little waboom is still on its rocky outcrop, and there
are still restios in the little stream, but all those majestic Clanwilliam
cedars I once walked past are long gone. Now my wife Nicky and
I walk the same trail as part of a large group of people who are here as
volunteers to plant nursery-grown seedlings into the wild. We now
know the altitude and terrain best suited for these trees' survival and
the focus is on putting as many trees as possible into these locations
before the wet season, to give them the highest possible chance of
establishing.

Over the years I have been on several of these excursions, and
I live in hope that we are making a difference. This time I get as far as
the neck between Tafelberg and Langberg and look down at the De Bos
site, but I don't go all the way to the trees as my knees are beginning to
complain. Well, that's old age for you! I really hope to visit them again.

. . .

4

. . .

KAURI

GRETEL BOSWIJK

SPECIES

New Zealand kauri, *Agathis australis* (D.Don) Lindl.

. . . .

LOCATION

Upper North Island Te Ika a Māui,
Aotearoa New Zealand

. . . .

ESTIMATED AGE

Commonly up to 1,000 years and, rarely, up to 2,000 years

The size and grandeur of New Zealand kauri have captured
the imaginations of people since the first arrivals in the
archipelago. Found naturally only in the upper North Island
Te Ika a Māui, kauri are the southernmost members of the
Agathis family and the largest trees by volume in Aotearoa
New Zealand. The most notable, Tane Mahuta, is over 16 ft
(nearly 5 m) in diameter, nearly 150 ft (over 45 m) tall, and
may be well over 2,000 years old. With British colonization,
kauri became economically important, with its timber used
widely in Oceania. Over time, the ecological, heritage, and
scientific values of these majestic trees have grown.

Author's note: Te Reo Māori is an official language
of Aotearoa New Zealand. Māori names and terms have
been used in this text in conjunction with English names
or translations where appropriate.

In Māori mythology, kauri is a descendant of the Supreme
Being, Io Matua Kore, who gives the tree its mauri (life force).
It is an embodiment of Tāne, god of forest, birds, and people.

I t is spring 1999, and I am on my first kauri tree coring expedition. I am a newbie at Auckland University, working as a research fellow on a palaeoclimate project with ecologist John Ogden and dendroclimatologist Anthony Fowler. John and I, along with my late husband Tony, are headed for Manaia Sanctuary, a closed conservation forest containing mature kauri in Te Tara-o-te-ika-a-Māui Coromandel Peninsula. Granted permission to visit, our objective is to add to a set of kauri tree cores collected in 1982. Then, using both sets, I will try to develop a tree-ring chronology for palaeoclimate reconstruction.

The drive from Tāmaki Makaurau Auckland takes us from city to wilderness as we cross the Hauraki Plain, follow the state highway hugging the peninsula's west coast, and wind our way up into the mountain range, all the way to the end of a private gravel road high on a hill. The views westward across the Hauraki Gulf to the distant city are magnificent, and I feel both lucky to be in this place and nervous about the work to come.

From here, we set off on foot and, trusting to John's memory, plunge off the ridge crest into dense forest, following possum trap-line markers downslope to the Kākātarahae stream. We scramble across the creek and up a north-facing slope into, quite suddenly, a stand of kauri.

They are impressive, with their thick, gray-pink cylindrical trunks rising up through the canopy. Some of these trees are more than

6½ ft (2 m) in diameter. On the edge of the stand, we find a huge tree, about 10 ft (well over 3 m) across, maybe four trees fused into one, damaged with the loss of limbs. Is this the tree called Tānenui (Great God of the Forest) or another big, unnamed, tree? We spend two days at Manaia, sampling young and middle-aged kauri. To our amazement, the clonk-clonk of the corer cutting through the wood fiber attracts a cacophony of kākā (*Nestor meridionalis*, a large native parrot), which become raucous in the branches above us.

Found nowhere else in the world

Like the kākā, these kauri are unique to Aotearoa New Zealand. Other *Agathis* occur elsewhere in the western Pacific (sometimes also called "kauri"). However, *Agathis australis* is found naturally only on a narrow strip of land and nearshore islands spanning less than 4° of latitude, from 34°23'S to 38°07'S. This encompasses Te Tai Tokerau Northland, Tāmaki Makaurau Auckland, and part of Waikato in the upper North Island Te Ika a Māui. Today, mature kauri are found in the fragmented old-growth forests with younger "second-crop" trees in regenerating forests. Kauri timber from trees felled during the late 1800s is found in buildings, boats, and furniture throughout the country. Kauri tree remains are preserved in lowland wetlands across the region, legacies of ancient forests from a thousand to tens of thousands of years ago.

   Agathis is a member of the Araucariaceae, a family of coniferous trees of Gondwanan lineage. About 83 MYA, Te Riu-a-Māui Zealandia, drifted away from the eastern margin of Gondwana, and by about 66 MYA had become separated from the other landmasses by ocean. Whether *Agathis* has been present since this time is debated, especially as around 90 percent of the landmass was submerged around 33 MYA. Leaf fossil evidence of *Agathis* is loosely dated to about 30–26 MYA, although whether this is *Agathis australis* is uncertain. The southern North Island Te Ika a Māui was submerged again around 5 to 2.5 MYA. Only in the last few million years has the familiar shape of Aotearoa New Zealand developed.

KAURI TIMELINE

• • •

Notable dates, climatic and cosmogenic
events recorded in the late Holocene
kauri record

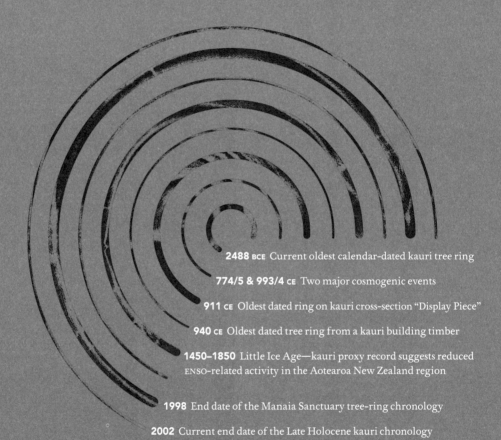

2488 BCE Current oldest calendar-dated kauri tree ring

774/5 & 993/4 CE Two major cosmogenic events

911 CE Oldest dated ring on kauri cross-section "Display Piece"

940 CE Oldest dated tree ring from a kauri building timber

1450–1850 Little Ice Age—kauri proxy record suggests reduced
ENSO-related activity in the Aotearoa New Zealand region

1998 End date of the Manaia Sanctuary tree-ring chronology

2002 Current end date of the Late Holocene kauri chronology

Macro-fossils identified as *A. australis* indicate the trees were present in southern North Island Te Ika a Māui from at least 350,000 years ago. In the most recent ice age, the range of kauri contracted to the area north of Tāmaki Makaurau Auckland, until about 12,000 years ago. As the climate warmed, the species range expanded again, reaching its current southern limit by about 1000 BCE. Why it did not expand further is debated. Is it climate, soil, seed dispersal and time to maturity, competition from other species, or some combination of these? We do know that kauri will grow (and set seed) further south, as evidenced by planted trees in public parks and private plantations throughout the country.

Within its current natural range, kauri occurs mostly in oceanic temperate forest below about 1,300 ft (400 m), with rarer stands at higher altitudes at the northern end of Te Tara-o-te-ika-a-Māui Coromandel Peninsula, and on Aotea Great Barrier Island. Although we talk about "kauri forest," the northern forests are typically characterized by broadleaf tree species, particularly taraire (*Beilschiemedia taraire*) along with ferns, nikau (*Rhopalostylis sapida*, a native palm), lianas, and large grasses. Some of the latter have sharp, cutty edges, treacherous when stumbling through the undergrowth in search of suitable trees. Further dangers await high up in the old trees, where large clumps of epiphytes nestle in the crooks of the branches. These epiphytes can sometimes become loose and fall, earning them the name "widow maker."

Kauri trees are often found on infertile, dry ridges, and sampling expeditions, including to Manaia Sanctuary, targeted trees on north-facing slopes because they are considered to have a stronger climate signal. (South of the equator, the sunny side is north.) But the preserved wood shows the trees once grew on or near lowland wetlands.

At a landscape scale, the occurrence of mature kauri—cohorts of trees around 600 to 1,000 years old—can be patchy. Small gaps of varying size may form as individual trees age and die. At a larger scale, regeneration is thought to be associated with catastrophic events such as cyclonic storms or fire creating big gaps in the forest. Following such catastrophes, dense cohorts of young kauri may establish under a nursery cover of manuka and kanuka, eventually breaking through the canopy. Today, this pattern of similar-aged cohorts can be seen in areas recovering from the land clearances of the last two hundred years.

Enduring appeal

In 1859, the writer Arthur Thomson described kauri as "celebrated and beautiful." What constitutes tree beauty? Kauri are not delicate-looking, like southern beech (*Fuscospora* sp.), or feathery like rimu (*Dacrydium cupressinum*). The honor of tallest tree goes to kahikatea (*Dacrycarpus dacrydioides*). Rather, kauri have a sturdy magnificence that can be seen as beauty—mature kauri are big, strong, and solid. Below ground, they hold themselves in place with deep peg roots. Close to the ground surface are fine feeding roots, protected by an accumulation of leaf litter, twigs, bark flakes, and humus that collects around the trunk. Too much trampling around the base of a tree can damage these roots, affecting tree health. The bark has a hammered appearance and is shed in large flakes, the fresh surface revealed in soft red hues. The leaves are thick, oval, and olive-green. The female cones are bright green, exploding with seeds as they dry.

Juvenile trees have a conical form but over time the lower branches fall off, leaving a straight trunk. Imagine many close-set poles with tufty tops. Eventually, some trees will capture space above the forest canopy, developing a thick, columnar trunk with upward-facing branches and a spreading crown. Others fare less well, dying back. As a result, mature kauri forest has fewer, more dispersed, and larger trees. Left to their own devices, these big trees will live for hundreds to thousands of years. Two of the largest trees we sampled at Manaia Sanctuary were more than 700 years old, while in Te Tai Tokerau Northland, the Waipoua Forest giants Tāne Mahuta and Te Matua Ngahere are likely well over 2,000 years old.

Human arrival

Kauri endured across the landscape for millennia, untroubled by much except storms, fire, or other cataclysmic events. Their forests were the domain of birds, bats, reptiles, frogs, and invertebrates, with fish and eels in the waterways. This changed around 1250 CE with the arrival and settlement of Māori from Polynesia. Forest gave way to gardens for growing kumara (sweet potato). At Waipoua Forest, repeated burning of the bush created large clearings used for hunting kiwi. Large kauri were selectively felled to make waka (canoes) of different sizes. The wood was also employed to make tools, while kauri resin was chewed like gum and the soot used in tattooing.

Pakeha (non-Māori) encountered Aotearoa briefly in the seventeenth century, then again in the late eighteenth century. Permanent settlements were established from 1814 onward and in 1840 New Zealand became a British Crown Colony. At the time, there was approximately 1.2 million hectares (3 million acres) of kauri forest in inland and rugged areas. As Pakeha immigration increased, demand for land and timber also rose. Within only about one hundred years, the once extensive forests were reduced to isolated fragments by a combination of logging and fire. Today only about 5 percent of the pre-1840 forest remains, with mature kauri forest encompassing 29 sq miles (75 sq km) mostly protected in the conservation estate.

This colonial history explains kauri as "celebrated and beautiful" —but it's not just about looks, it's about the timber yield. During the 1800s, reports of the big trees spread beyond Aotearoa New Zealand through the writings of late Georgian- and Victorian-era explorers, sojourners, missionaries, botanists, and other scientists. They described the tree, its leaves, cones, and seeds, and the forest environment. Its timber and economic utility were presented in glowing terms:

> "*A. australis* or New Zealand Cowdie Pine, is one of the finest trees in the world, often growing perfectly straight to the height of 100 feet or more, and yielding one of the best descriptions of wood for masts." — J.C. Loudon, *An Encyclopaedia of Plants*

Kauri wood and furniture was shown abroad, including at the London International Exhibition of 1862, in the Economic Museum at Kew Gardens, and at the 1886 Colonial and Indian Exhibition, London. Seeds were collected and dispatched to Kew and other botanic gardens across the British Empire, and trees were raised for display. The potential yield and profit of these trees was described in breathless terms: "A Kauri-tree, only forty feet high and thirty-seven in circumference, when sawn up, yielded 22,000 feet of good timber, which was sold for 500l, leaving a clear profit to the owner of 300l" (*Chatterbox*, 1886, London).

An incredible number of trees were cut down and turned into buildings and boats, furniture, and even street paving blocks. It is mind-boggling how fast the forest went and how much was lost—not just the trees, but the birds, bats, reptiles, insects, and water creatures.

Studying tree rings

Late nineteenth- and early twentieth-century botanists and forest scientists certainly looked at the trees and timber, mostly counting kauri tree rings to estimate growth rates and consider production forestry potential. In the 1950s and '70s, U.S. researchers collected tree cores and attempted to build tree-ring chronologies for climate reconstruction and archaeological dating. Success was variable, as the tree species were found to be challenging. Additionally, swamp kauri "biscuits" (cross-sections) were collected for radiocarbon research, although no tree-ring chronologies were built.

The 1980s saw a flurry of local activity. John Ogden set up the first tree-ring laboratory at the University of Auckland and, with his students Moinuddin Ahmed and Jonathan Palmer, along with Anthony Fowler, researched kauri ecology, palaeoecology, and dendroclimatology. They collected tree cores from Northland to Waikato, constructed site chronologies, and developed foundational knowledge about kauri dendroclimatology. John and his students established that kauri are sensitive to climate, recording changes in the climate in the year-to-year changes of their growth rings. A visiting post-doc, Martin Bridge, built the first swamp kauri chronology, positioned in time by radiocarbon dates.

El Niño and the kauri

Another push occurred from the mid-1990s onward, when Anthony Fowler identified a statistical relationship between kauri tree growth and the El Niño–Southern Oscillation (ENSO), a coupled ocean-atmosphere climate phenomenon centered on the tropical Pacific. It manifests as periodic warming (El Niño) or cooling (La Niña) of sea surface temperatures in the eastern Pacific Ocean and changes in trade wind strength and direction (the Southern Oscillation component). These events impact global climate and are connected to extreme weather events such as storms, floods, and droughts. Although New Zealand is not in the core region, its climate is affected by ENSO events, with cooler and drier conditions in the North Island during El Niños and warmer-wetter conditions during La Niñas.

ENSO events tend to develop in the latter part of the year and last several months, coinciding with the kauri growth season. In very simple terms, kauri trees tend to grow very well in cool-dry conditions (El Niño) and less well in warm-wet years (La Niña). This relationship, and the longevity of kauri, raised prospects of a long proxy record of ENSO activity from the southwest Pacific. Since ENSO can have significant environmental impacts affecting human society, such a record would contribute to our understanding of the frequency of past events and how this might change in a warming world. It was at this time, in late 1999, that I joined the Auckland University team.

To advance this work, we needed to bring the modern record forward from the 1980s, capturing more recent ENSO events. We also needed to push our record back in time as far as we could go, allowing a long view of ENSO activity. Having more dated series would improve the quality of a master chronology—an average record of all crossdated series—for the reconstruction. The Manaia Sanctuary trip was the first of many expeditions to core living trees, sample kauri timbers from buildings, and collect swamp kauri biscuits. And it was the start of over a decade of chronology building for me, carefully piecing together tree-ring series into what eventually became a continuous 4,500-year-long record of kauri tree growth.

Coming to terms with kauri

Although born in Aotearoa New Zealand, I studied archaeology at university in England and, for my honours dissertation, recorded and tree-ring dated a large, late-sixteenth-century oak timber-framed barn. This led to a PhD using dendrochronology to reconstruct a buried forest and, on the side, training in the analysis of archaeological wood from the dendrochronology team at Sheffield University. When the post-doctoral research position in dendrochronology arose in Auckland with John Ogden and Anthony Fowler, my husband Tony and I decided to take up the opportunity.

When I started, my task list included updating the modern sites of standing kauri, which is what took us to Manaia Sanctuary. That site yielded my first kauri chronology: 730 years, spanning 1269 to 1998 CE, still the longest modern site chronology. In my first year, I reviewed and updated other modern kauri sites to build a new master chronology.

I also ended up in the attic of a duplex, "Sinton Road," sampling roof timbers to help establish a construction date before it was demolished for a roading project. The site chronology dated to 1711–1903 CE, and, excitingly, two timbers still had the outermost growth ring. They came from a tree (or trees) cut down in late 1903 or early 1904, supporting other evidence suggesting that the structure was built by early 1905. Sinton Road helped us learn about the potential of dendrochronology to contribute to the study of New Zealand buildings and how building timbers could also be useful for the palaeoclimate research.

At the end of 2001, Anthony Fowler secured funding for a new three-year project furthering the kauri-ENSO research. I would lead development of tree-ring chronologies, extending the modern record back in time and building swamp kauri chronologies. My husband Tony was diagnosed with terminal cancer not long after the project began— losing myself in the wood cells and the routine, steady measurement of tree rings was one way to manage at a time of great personal sadness.

Fitting the "strips" together

There are many different ways to describe the process of building a tree-ring chronology. It's like doing a jigsaw! But here is a different image. My mother is a noted ceramic artist. In her practice, she would hand-coil strips of clay to make large sculptural works. Making a chronology is similar but linear and numerical rather than enclosing space. Our "strips" are ring-width series that are lined up against each other, with only one place in time that they can go. Clay and wood are tactile, but of course we are working with tree-growth patterns expressed as numbers. The "art" becomes visible as graphs, with an average chronology show-ing the annual ups or downs of multiple trees.

Between 2003 and 2006, the chronology building was intensive. Many hours were spent at the microscope and traveling stage, measur-ing the ring widths and running the computer programs to identify statistical matches between ring-width series. It wasn't easy. To mangle Longfellow: When kauri are good, they are very good indeed, but when kauri are bad, they are horrid! What this means is that, sometimes, a series will crossdate like a dream. But sometimes a series might have locally missing rings messing up the sequence. Or there may be false ring boundaries caused by a change in conditions during the growing season, making one ring look like two. Every series was printed as a paper graph and I'd stand at the light table, pencil in hand, overlapping sheets, comparing patterns, and going back to the wood samples hunting errant rings. Check, check, check. It takes time and patience to work out such problems, but is very satisfying.

The modern kauri record was updated to 2002—a sequence from a large display cross-section pushed the record back to 911 CE and new chronologies from buildings were added. Floating chronologies (not calendar dated) from the swamp kauri were built and linked up. The ultimate goal, finding the overlap between the calendar-dated and floating records, was a quiet achievement in the month or so before Tony and I went to my family for his last few weeks. I left the data set with Anthony Fowler to check. Coming back and picking up the pieces again meant being able to put calendar dates on the kauri record as far back as 1724 BCE. Since then, more building timbers and swamp kauri samples have been added to the data set. Our late Holocene kauri master chronology goes further back now, to 2488 BCE, with sections that are older and still floating in time. And of course, this was a collaborative effort with different team members making contributions along the way.

Making use of the long kauri record

How has this calendar-dated kauri record been used? Well, it contains multiple stories about climatic, environmental, and social change in the southwest Pacific. Here, I will highlight three relating to ENSO, radio-carbon calibration, and our attempts to date Māori artifacts.

Of most importance—because it underpinned the chronology building—is the 700-year proxy record of ENSO variability developed by Anthony Fowler. He observed a reduction in ENSO activity in the Aotea-roa New Zealand region during a cool period known as the Little Ice

Age, followed by rising activity as global temperatures increased from the mid-1800s. The findings suggest that with a warming world, ENSO activity will increase, and Aotearoa New Zealand's climate will become more dominated by these events. Such change could produce more intense ENSO-related drought or rainfall events, impacting people, their homes and livelihoods in rural and urban areas.

The second important contribution is the use of kauri to build carbon-14 (C14) chronologies, which can accurately position environmental and societal events and changes in time, as well as objects. C14 can be measured from blocks of known-age rings or even single years. Kauri wood from the past 3,000 years has gone to colleagues building radiocarbon chronologies, including the 2020 iteration of the Southern Hemisphere Radiocarbon Calibration Curve, and has also been used to document extreme cosmogenic events in 774/5 and 993/4 CE.

I have also worked on more nineteenth- and twentieth-century buildings to help identify when they were built or modified, contributing to understanding of the structures and the people associated with them. There have been attempts to date Māori artifacts, but here, success has been evasive. Cultural considerations are important—is it appropriate to core sample a taonga (treasure) like a canoe? Granted permission from iwi (tribes) leaders, we have tried to date two canoes, but encountered practical problems of sampling and insufficient tree rings. Yet, from failures come new opportunities. Researchers in Britain have developed a new method of dating using the annual pattern of the stable oxygen isotope that is preserved in tree rings. These can be crossdated between trees, just like ring widths. We are now exploring if this method could offer an alternative approach for dating archaeological kauri wood.

Window into an ancient world

I have stayed in the shallow end of time, working with kauri from the past 5,000 years or so, but colleagues have studied much older kauri. There is swamp kauri from around 11,000 to 12,000 years ago that was alive during the Younger Dryas, an abrupt cold snap at the start of the

Holocene. And then there is the "Ancient Kauri," the preserved remains of trees that lived sometime between 27,000 years ago and 60,000 years ago, during a period known as "Marine Oxygen Isotope Stage 3." These trees can be very, very big, more than 13 ft (4 m) in diameter, and very old. Cross-sections have been recovered that have more than 2,000 rings. Such ancient wood is globally rare and therefore has high scientific value. The tree-ring and radiocarbon chronologies built from this wood span several thousand years and provide valuable insights into global and hemispheric-level environmental and climatic changes.

Most notable among this group is the "Ngāwhā kauri," a large tree recovered from the far north of Te Tai Tokerau Northland. This tree was alive some 41,000 years ago and lived through a magnetic reversal of the poles, called the Laschamps excursion. Stored within its wood is a record of changes in atmospheric radiocarbon before and during the reversal. A 1,700-year C14 chronology has been built from this tree and a few others of similar age. Aligned with other long-term proxy climate records, the Ngāwhā record illustrates environmental changes associated with a weakened magnetic field before and during the reversal for the mid to low latitudes of both hemispheres. While some of the interpretations about the environmental and social impacts are contested, the work stands testament to the great value of the ancient kauri as a window into a long past world.

Our shifting connections with kauri

Working with kauri—handling many, many pieces of wood from trees, buildings, and swamps, and developing the long chronology—pushed me to grow my knowledge of the shifting relationships that humans have with this species.

In Te Ao Māori (the Māori world) the native totara tree is culturally more important because it is found throughout Aotearoa New Zealand and was used widely for waka and other objects. The proverb *"Kua hinga he totara i te wao nui a Tane"* ("A totara has fallen in the forest of Tane") acknowledges that a person of great importance has died. Yet, kauri is a national taonga (treasure). In modern Māori philosophy, kauri has spiritual qualities as a descendant of the Supreme Being, Io Matua Kore, who gives the tree its mauri (life force). Kauri is an embodiment of Tāne, atua (a god) of forest, birds, and people. In northern versions of the Māori creation myth, Tāne grows like a kauri to separate his earth mother, Papa-tū-ā-nuku, and sky father, Rangi-nui, to bring light into the world.

European contact and nineteenth-century settler colonization insti-gated a different relationship to kauri. Pakeha were awed by the size of the trees. Descriptions of cathedral-like groves occur in print and there are romantic depictions of kauri in art. Even so, pragmatism and eco-nomic concerns won the day. The trees were considered a "God-given" resource to be used and kauri gum was economically valuable, exported to Britain and America to make an early form of linoleum and for var-nish. Ancient gum was dug out of the ground while living trees were repeatedly slashed to make them bleed. This happened to some kauri at Manaia Sanctuary, causing deep wounds to form on the trunks, possibly affecting the tree-growth patterns captured in the 1980s tree cores.

Kauri remain "nowhere else to be seen"

From the early 1900s, exploitation gave way to concern for the survival of the remaining kauri forest and trees. Advocates for protection emphasized the ecological uniqueness, grandeur, scenic beauty, and antiquity of the species and its forest environment. Kauri became part of the country's national heritage and preservation became a moral duty. The first formally protected remnant of kauri was Trounson Kauri Park in Te Tai Tokerau Northland (opened in 1921). In the 1950s, a national petition resulted in designation of part of Waipoua Forest as a Sanctuary. Manaia Sanctuary sits within the wider Coromandel Forest Park, proclaimed in 1972 when the Forest Service shifted toward management for conservation. These events fit within a wider, growing environmental movement in Aotearoa New Zealand during the twenti-eth century, fighting for the cessation of logging and protection of native forests and their fauna.

Some of the protected old-growth forests are open to the public. Trounson Kauri Park has evolved into a "Mainland Island" with intensive predator control around the perimeter to enable native birds, insects, reptiles, and aquatic life to thrive within its confines. Take a night walk in the park and you might hear or even see rare kiwi (*Apteryx* sp.) birds. At Waipoua, the forest giants Tāne Mahuta and Te Mātua Ngahere are just a short walk away from the state highway, while longer trails pass by other notable trees. Other kauri forests are inaccessible without a permit, protecting the trees and forest from further human-induced harm.

Naturally regenerating kauri forest is found on public and private land. And kauri has been, and is being, planted throughout the country by community trusts, conservation organizations, and private individuals. Within its natural range, the goals are to restore and recreate kauri forest, often combined with pest and predator control, to enable native species to flourish and restore ecosystem linkages. Further south, trees have been planted in botanic gardens for display, or in stands to study potential timber yield. In the lower South Island Te Waipounamu, a new scheme is even reforesting farmland with kauri to capture and store carbon in the wood.

These different plantings are all long-term projects; in about five hundred years' time, all being well, the kauri will be mature, and the forests spectacular. I sometimes wonder if future people will remember who made these plantings and why, or if they will be viewed simply as natural places full of useful timber trees.

"All being well"—there's the catch. Kauri face challenges. At present, the most pressing issue is a deadly soil-based pathogen, *Phytophthora agathidicida* (PTA), which is causing kauri dieback throughout its natural range. People and animals are carriers of PTA, which can be transported in tiny amounts of soil on shoes, equipment, vehicles, or the feet of animals. Trees are infected through their roots, with sick trees developing basal lesions that bleed sap and/or suffer yellowing of leaves and thinning of the canopy, eventually leading to branch and then tree death. The national Tiakina Kauri (Kauri Protection) program aims to limit the spread of the pathogen through awareness raising, education, and practical measures.

It is now routine and expected, when visiting accessible kauri forest, to use the shoe cleaning stations at the entry and exit points to minimize possible spread of the pathogen. Walking tracks have been replaced with raised boardwalks to prevent trampling on the forest floor and around the sensitive kauri roots. Rāhui (restrictions) have been placed on some forests, closing tracks, or even whole forest blocks. In 2016, Manaia Sanctuary was completely closed to public access to help maintain its PTA-free status.

An enduring tree, a last picture

During our trip to Manaia, Tony took a photograph of John Ogden at the base of the big tree that might be Tānenui. In this picture, John is diminutive compared to the thick trunk and branches that rise high above him, the gaps from the broken limbs prominent, along with the creepers growing up the trunk and the epiphytes in the crooks of the boughs.

Studying that picture again reminds me of the magnificence of these trees and their capacity to endure. I have not been back to the Sanctuary since, so do not know how that tree is faring. I am hopeful it is still standing, keeping company with the kāka that are, no doubt, still being raucous in the forest canopy.

• • •

5

...

BALD CYPRESS

MATTHEW THERRELL

SPECIES

Bald cypress, *Taxodium* spp. (L.) Rich.

....

LOCATION

Southeastern USA, Mexico, and Guatemala

....

ESTIMATED AGE

Over 2,600 years old

The bald cypress, *Taxodium* spp. (L.) Rich., grows throughout much of the southeastern United States in the swamp forests of the region as well as along riverbanks from southern Texas south to Guatemala. The oldest-known individual grows along the Black River in North Carolina and is more than 2,624 years old. Bald cypress trees in Mexico have been documented at over 1,000 years old and trees over 500 years old can be found throughout its range. Bald cypress can also grow to immense size, easily reaching over 100 ft (30 m) in height and 12 ft (nearly 4 m) in diameter, with swollen "knees" or buttresses that can be up to 10 ft (3 m) high. Bald cypress trees produce large quantities of attractive pale-brown to reddish wood that is resistant to decay and very desirable for construction. In the late nineteenth and early twentieth centuries, commercial exploitation boomed in the southeastern United States, leading to excessive logging.

*The bald cypress's sensitivity to drought, along with
the fact that it is so long-lived, makes it an ideal candidate
for tree-ring studies.*

oday, only a few large, protected tracts of old-growth cypress remain. Bald cypress tree-ring records have been used to link the failure of early English settlements on the East Coast, such as the Lost Colony of Roanoke and Jamestown's Starving Time, to some of the worst droughts in the region over the past millennium. There are three recognized species in the genus of bald cypress, though only a botanist would likely notice the minor differences among them. The first, *Taxodium distichum*, what most people call bald cypress, or "swamp cypress," grows from roughly southern New Jersey, down the East Coast to Florida, across the Gulf Coast to Texas, and inland to southern Illinois and Indiana in the Ohio and Mississippi River Valley region. A generally smaller cousin, pond cypress (*T. ascendens*), can be found in the far southeastern portions of this same range, particularly in small natural ponds in the coastal plain. A third species, *T. mucronatum*, or "Montezuma cypress" in English, which is also essentially indistinguishable to laypeople, grows from Texas down through Mexico into Guatemala. This genus of trees has a very ancient lineage, being members of the Cupressaceae family, which evolved in the Jurassic (201–145 MYA) or earlier and includes many other species of ancient trees such as the Californian redwoods (*Sequoia sempervirens*) and giant sequoias (*Sequoiadendron giganteum*), or the alerce (*Fitzroya cupressoides*) in South America.

MATTHEW THERRELL

The wood of bald cypress is highly valued for its rot resistance, light weight, and easy workability. In North America, Native peoples, for example, the Seminole in Florida, used the wood for dugout canoes and a variety of other uses, including in ceremonial contexts. Larger-scale harvesting of bald cypress for construction began during the Colonial Period, around 1718, particularly in French Louisiana, and reached its apex in the early to mid-twentieth century when many millions of board feet were harvested annually, particularly in the Carolinas, Florida, and Louisiana. Often, there were stands and individual trees that were spared from cutting because they were hollow and/or misshapen. Today, very few extensive stands of uncut bald cypress still survive, but smaller remnants of ancient uncut bald cypress can be found all over its range. Some of the most well-known preserves are the Audubon Society preserves at Corkscrew Swamp in Florida, and the Francis Beidler Forest in South Carolina. Outside the few reserves, hardly any commercially valuable timber trees were left.

Getting to know the bald cypress

Although I was familiar with bald cypress growing up—I often fished in and explored my local swamps, which included the phenomenally beautiful National Natural Landmark of the Mobile-Tensaw Delta—it was not until I went to graduate school and began studying dendrochronology that I began to know and love the truly fascinating story of this tree.

My first scientific experience with this amazing tree was in the winter of 1996 in northeastern Mexico, when I went with my PhD advisor Dr. David Stahle at the University of Arkansas, one of the world's foremost experts in dendrochronology, to sample bald cypress trees in Tamaulipas, Mexico. For those who remember, this was the year that comet Hyakutake made its spectacular appearance. Seeing this comet in the pitch-black skies over the mountains of Tamaulipas the night before our planned sampling was one of the most awe-inspiring experiences in my life and it seemed like an auspicious sign for our work. Another

BALD CYPRESS TIMELINE

• • •

Climatic events evidenced by the bald
cypress tree-ring record

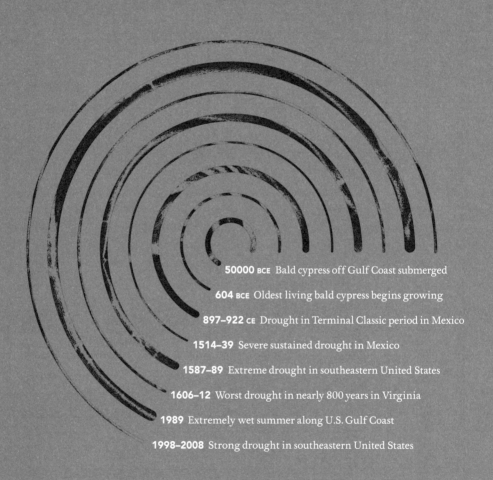

50000 BCE Bald cypress off Gulf Coast submerged

604 BCE Oldest living bald cypress begins growing

897–922 CE Drought in Terminal Classic period in Mexico

1514–39 Severe sustained drought in Mexico

1587–89 Extreme drought in southeastern United States

1606–12 Worst drought in nearly 800 years in Virginia

1989 Extremely wet summer along U.S. Gulf Coast

1998–2008 Strong drought in southeastern United States

Although we would not analyze the samples we collected in Mexico right away, my experience with these magnificent trees was quite profound.

fascinating surprise for me was that, unlike the swampy setting where one would usually find bald cypress in the eastern United States, in Mexico and the drier parts of Texas, it is found growing right along the banks of the typically smaller rivers and streams in these more arid regions. At our site in Tamaulipas, called Rio Sabinas, or Cypress River in English, the floodplain is rocky; the streams, which are often fed by crystal-clear springs boiling up from sinkholes in the limestone called cenotes, can become raging torrents when heavy rains fall higher up in the mountains. Not long before our visit a hurricane had struck this region and the floodwaters had scattered all sorts of debris including entire trees across the floodplain and against the huge bald cypress trees we intended to sample. This, too, was very unlike the slow-moving floods I had seen in the coastal swamps I was familiar with.

In addition to the difficult terrain, coring bald cypress is more difficult than most tree species because we need to use ladders or climbing gear to take samples above a tree's large, buttressed base, where the rings are more clearly formed. In addition, there are the many beautiful and fascinating—though sometimes annoying (poison ivy) or attention-grabbing (snakes, huge spiders)—neighbors we routinely encounter. On this particular visit, we even had a run-in with killer bees, which made it particularly memorable. Although we would not analyze the samples we collected in Mexico right away, my experience with these magnificent trees was quite profound. Not only were they far more massive than any bald cypress I had seen before, they were also the oldest trees I had ever encountered. I quickly realized how useful such ancient trees would be for our research and why we had come so far to make our collection. This growing appreciation of bald cypress trees, married with an interest in history, would very soon help set the course of my scientific research for the next three decades.

Using dendrochronology to help understand the past

I recall vividly the day an archaeologist from Jamestown visited our team at the University of Arkansas Tree Ring Lab to discuss the possibility of using tree rings to study the early history of the colony. Although already five years into my graduate training as a scientist, I had previously earned a B.A. in history, and I was immediately fascinated by the notion that I could be a part of such an important historical story.

History buffs may know that the settlers of Jamestown, Virginia, who arrived in 1607, bear the distinction of founding the first successful English colony in North America, and many will know that Jamestown also came very close to joining the list of failed colonization efforts such as the mysterious Lost Colony of Roanoke, Virginia, only a few decades before. The incredibly tragic story of Jamestown's near failure was well documented by historians, but most had attributed the suffering to inept colonists. Some, including our guest, however, speculated that climate had perhaps played a role. By an amazing coincidence, ancient bald cypress trees growing very near Jamestown along the Blackwater and Nottoway Rivers in Virginia did in fact record the stunningly difficult circumstances that the Lost Colony and Jamestown colonists endured during their attempts at settlement. And while there is very little historical documentation describing the impacts of the drought on Native peoples in the region, including the local Powhatan, what few records exist along with the tree-ring data seem to clearly indicate that they also faced extreme hardship as a result.

This unexpected visit proved the perfect connection between history and tree-ring research. Certainly Dr. David Stahle was well known as the person to talk to about bald cypress tree rings. Indeed, Dave and I, along with his other students and scientific collaborators, have developed scores of bald cypress tree-ring records all over the species' range and used these records in multiple groundbreaking studies of climate focusing in particular on drought and associated large-scale climate patterns, and ecology. His intimate knowledge of these ancient trees has also led him to become one of the most dedicated spokespersons for conserving what few stands of ancient bald cypress trees still exist.

Dave was really the first tree-ring scientist to recognize the enormous value of the species to reconstruct how climate variables such as rainfall or streamflow change over time. Most every other dendrochronologist assumed that a tree growing in the water would not have any sort of sensitivity to rainfall or drought conditions. Yet bald cypress is extremely sensitive to even fairly minor reductions in its water supply. This sensitivity to drought, along with the fact that bald cypress lives much longer than any other tree species in the eastern United States, makes it an ideal candidate for tree-ring studies. Dave's chronologies near the settlement would provide an excellent test of the notion that drought may have contributed to the near failure of Jamestown.

The only drawback to using bald cypress for tree-ring studies, besides the difficulty of sampling the frequently enormous trees, is that, because they are often so old, the actual tree rings can be extremely narrow. Even under the microscope, a drought year perhaps represents only a few cells of growth, and is barely discernable. To make matters worse, these "micro" rings are often not evenly distributed around the full circumference of the tree. Fortunately, by taking samples from many trees at a site, we are able to crossdate every single ring from each tree sampled to verify the exact year that it was formed. There is no "plus or minus" in our dating. This is what makes dendrochronology the most exacting of all the geochronological techniques, and this is what makes it possible to shed light on the climate or other environmental impacts that may have affected people such as those early settlers at Jamestown.

Of course, we had no idea at the time that the ancient trees we studied had been witness to the extreme droughts that contributed to the stark suffering at Jamestown and among native populations as well as seemingly the mysterious disappearance of the Lost Colony. But our findings did reveal that, specifically, the tree-ring record of climate variability in Virginia shows that the Lost Colony and Jamestown settlers arrived during two of the worst droughts in roughly 800 years. We described our findings in a paper published in the journal *Science* and it was picked up by scores of media outlets including *The New York Times*, which ran the story on the front page. Even *National Geographic* sent a photographer to our lab to document our findings. For those of you who grew up in the heyday of this magazine you can perhaps imagine how impressive that was to someone like me.

Discovering Mexico's oldest and most venerated trees

All the publicity surrounding this article was exciting and made me want to do more research on how past climate had affected the environment and society. At the time, there were relatively few tree-ring records of climate in Mexico, particularly in the tropical regions, and it has abundant historical and archaeological records of climate impacts on its people. Along with other dendrochronologists from the United States and Mexico, I spent the next decade or so making new tree-ring records of climate all over Mexico. Along the way, we found many of the oldest-known and most amazing trees in Mexico.

Mexico has an incredible diversity of tree species—for example, it has more species of oak and pine than anywhere else on Earth. Yet I think that if you asked most people there, they would agree that bald cypress, often referred to as its Spanish name of sabino, or ahuehuete, as it is called by Nahuatl-speaking people such as the Aztec, or Mexica (as they called themselves), is probably the country's most culturally important tree species. Ahuehuete basically translates to "Old one of the river" and is known as the Tree of Life in Nahua cosmology. Its popular status was legally defined in the early twentieth century when ahuehuete was proclaimed the national tree of Mexico.

In many locales, these trees are venerated if not outright deified. Famous and celebrated individual bald cypress trees can be found all over the country. Without a doubt one of the most famous trees in the world is the so-called Tule Tree or, more formally, the Santa Maria del Tule Tree, found in Oaxaca, Mexico. It is the "fattest" (that is, the trunk has the greatest circumference) of any tree known in the world.

Its circumference is a bit over 130 ft (40 m), making it slightly greater than California's well-known General Grant giant sequoia. The Tule Tree has been estimated to be somewhere between 1,500 and 6,000 years old! As a professional dendrochronologist I think the notion that this tree is around and perhaps over 2,000 years old is reasonable. This tree was so culturally significant at the time of Spanish "conquest" that a church was constructed next to it, no doubt to co-opt some of the tremendous regard in which it was held by Indigenous peoples in colonial times. The Tule Tree is still an incredibly popular destination, visited by thousands of people annually from Mexico and all over the world. It was at one time considered for listing as a UNESCO World Heritage site, which I certainly think it deserves.

Other famous examples of ahuehuete include the now dead Árbol de la Noche Victoriosa (Tree of the Victorious Night), which celebrates an Aztec military victory over Cortez in 1520, as well as scores of bald cypress trees said to have been planted by Mexican rulers, in Mexico City's Chapultepec Park, roughly five hundred years ago. Chapultepec Park is in many ways analogous to New York City's Central Park; one of the most nerve-wracking experiences I ever had in the field was when we sampled these amazing trees while thousands of Chilangos (as residents of Mexico City call themselves) walked by and rightly wanted to know just exactly what we were up to. In one particularly funny case, one of the trees we were coring was partially hollow (as they often are) and had collected a huge amount of water inside the hollow trunk. While coring this tree, a great stream of water gushed from the hole the borer made. The water shot out maybe ten feet from the tree for 15 minutes or more, and this drew quite a crowd. One young man was even brave enough to taste the "mezcal de ahuehuete" as he called it—his reaction suggested he was dissatisfied, to say the least.

During our tree-ring research collecting in Mexico, we discovered many other amazing ahuehuete that are not so well known. Some of the most fascinating to me were growing in a beautiful marshy area called Los Peroles near the town of Rioverde in San Louis Potosi. This is one of the few places in Mexico where I have seen an extensive open grassy wetland, something like one might see along the Atlantic or Gulf Coasts, except in this case with crystal-clear streams and huge ancient ahuehuete scattered sparsely about. When we visited this site and collected samples, we discovered that one of the trees we cored was over 1,000 years old, making it the oldest documented tree in Mexico. To a trained dendrochronologist it is obviously a very old tree, having all the distinguishing characteristics of very old age. Much like it is very easy for most people to recognize elderly people, once you have seen hundreds or thousands of old trees, it is quite easy to recognize them. Some of the characteristics of old age in trees include a strongly twisted stem, few limbs that are typically large and heavy, and often the crown will be broken and dead. They tend to have very thick bark (relative to what is typical for the species) and very large buttressing at the ground. If one imagines all the violent storms, droughts, possibly floods, and even fires that a tree has to endure over a thousand or more years of life, it is easy to picture how bent and broken they may become. We ended up calling this tree Maximino to reference its great age and as a nod to Maximino Martinez, one of the most famous botanists in Mexico.

We found another ancient stand of bald cypress in a completely different environment, at the bottom of a steep, straight-walled canyon in the state of Queretaro, central Mexico, at a place called Barranca de Amealco. Here, the ahuehuete trees struggle to cling to the bedrock of the tumbled boulders along the canyon wall and those at the bottom of the canyon are sometimes tortured into fantastically twisted shapes by the violent torrents that flood the canyon in wet years. Here, too, we found trees over 1,000 years old.

Connecting old trees to Mesoamerican cultural history

Finding these trees and developing a tree-ring record of climate for over 1,000 years has long been a goal of dendrochronologists and archaeologists. The proximity of the Amealco trees to central Mexico's former Classic period (250–900 CE) Mesoamerican cultural centers (about 60 miles/100 km from the former Aztec capital of Tenochtitlan) allowed, for the first time, a precisely dated, year-by-year history of the impact of climate all the way from the Late Classic period (600–900 CE) up to modern times. This record supported earlier paleoclimate evidence from lake sediment and cave deposits (speleothems) of ninth- and tenth-century megadroughts, which others had suggested led to major social and political disruption of the Classic Period Maya. However, it also provided much new insight into the rise and fall of several other Mesoamerican cultures, including the Toltec and the Aztecs.

Using tree-ring records to examine how climate may have impacted the Aztec was of particular interest to me, because the Aztec produced some of the few written (actually, pictographic) records that had survived through time and the Spanish colonization. Comparison of these records with the tree-ring data showed that several of the most intense famines in Aztec history were clearly driven by drought and to some extent unseasonable frosts that may have both been related to volcanic events as far away as Indonesia and Iceland. I wrote my dissertation and, along with my colleagues, many other papers on these new insights into climate history—including its large-scale drivers such as El Niño—and some of the impacts on pre-Hispanic and colonial society, including famines and plagues that killed millions. Since then, my colleagues in Mexico have continued to find dozens of ancient ahuehuete sites throughout its range in Mexico and even into Guatemala, though none yet as dramatically old as those in Los Peroles and Amealco.

Secrets of extinct birds and underwater forests

Around the time that my colleagues and I were publishing our findings gleaned from the ancient bald cypress of Mexico, another mysterious and nearly forgotten denizen of the southern U.S. swamps made the front page of newspapers everywhere and focused the world's attention on the importance of the ancient bald cypress ecosystem. I mean, of course, the Great God Almighty Bird, as the ivory-billed woodpecker (*Campephilus principalis*) was once known. In 2004, a group of scientists, mostly from the renowned Cornell Laboratory of Ornithology, reported

MATTHEW THERRELL

the apparent rediscovery of these supposedly extinct birds in the Big Woods region of the Mississippi Delta in eastern Arkansas. Thought to be extinct since the late 1930s, the reported rediscovery of living birds in the bottomland swamps of the Cache River National Wildlife Refuge created a frenzy of news reports, and scores of scientific papers and a few excellent books were written describing the resulting (and ultimately unsuccessful) hunt for other members of this extinct bird. Fortunately, this mystery also encouraged interest in preserving what few remnants of ancient cypress swamps might still harbor this beautiful and mysterious creature thought to have been annihilated along with the millions of acres of virgin forests where the birds once found refuge.

Coincidentally, 2004 was also the year that Hurricane Ivan devastated the Gulf Coast of Alabama. Unbeknownst to scientists at the time, the storm's ferocious wave energy scoured the sea floor and uncovered the still-standing trunks of bald cypress trees that grew there more than 50,000 years ago. Huge standing trunks and fallen logs of trees lay scattered along an ancient river channel just as they were in life. These trees last grew in the late Pleistocene, 50 to 75 thousand years ago when sea level was as much as 200 ft (60 m) lower, but rapidly rising. The fascinating discovery of this so-called "underwater forest" caught the attention of scientists, including dendrochronologists who were able to take samples of the still-fragrant wood as well as the layers of peat and sand the trees once grew in. The scientists studying this ancient version of the modern Mobile-Tensaw Delta, referred to by some as America's Amazon, were surprised to find such ancient wood preserved at the bottom of the Gulf. Fortunately, the eventual burial of the forest by the rapidly rising ice-age Gulf had preserved the wood almost perfectly. Tree-ring studies of the samples brought to the surface showed that at least some of the trees were nearly five hundred years old and that all the sampled trees died in the same year. This clearly suggests that the site experienced some sort of extreme mass-mortality event, possibly a massive flood of the river they once grew alongside, which deposited enough silt and mud to kill and probably entomb the forest even before the sea eventually rose.

Although the discovery of the Underwater Forest, as it became known, was surprising to many, for those who spend time in coastal waters (including cypress swamps) a somewhat similar phenomenon may be routinely witnessed today. Any boater growing up along the Gulf or Atlantic Coasts will likely have seen standing dead trees or decaying stumps of bald cypress and other coastal species along the shore or even well out into the water of their favorite bay or estuary. One does not have to be a scientist to know that where these dead trees stand in water there once was land. In many coastal areas and a few inland spots, so-called "ghost forests" may be seen where a variety of environmental forces, from earthquakes to sea-level rise, have killed large swathes of coastal or riverine forest. The insidious creep of toxic salt-water intrusion resulting from human-caused sea-level rise is probably the most common driver of coastal tree mortality these days and, as the Underwater Forest demonstrates, this process has gone on for many millennia as sea levels rise and fall.

Bald cypress subfossil remarkably preserved

Most people would probably view a dead tree, especially a very ancient one, with at least some sense of sadness. But those of us who study tree rings are fortunate to know that even in death the environmental history preserved in a dead tree's rings can inform our understanding of the history of climate, sea level, and other extreme events as well as their impacts on the environment and people for, in some cases, more than 10,000 years.

In addition to being very long-lived, the fact that bald cypress is extremely decay-resistant means that when we go into the swamps to sample living trees, we often also find standing dead and fallen trees that have been dead for centuries. Moreover, as sites such the Underwater Forest demonstrate, when these logs are buried fairly rapidly, they can survive for tens of thousands of years. Finding this not modern but not old enough to be fossilized dead material, which we generally refer to as subfossil, has often amounted to literally tripping over them in the swamps. If a dead tree is still standing or they are not already buried too deeply, it is quite easy to use a chainsaw to cut a sample (something we would never do to a living tree). Collecting these subfossil samples has, in many cases, allowed us to develop tree-ring records much longer than would be possible with only living trees. As long as some outer portion of a dead sample overlaps a living tree in time (or another dead one that does), we can potentially create a chain of exactly dated samples going back well beyond the oldest living tree of the species.

In some cases, such as the Underwater Forest, or logs that have been dug up from deep sand quarries or during road construction, the logs are too old to connect to our current records—these are what we call "floating chronologies," which will have to wait until we are able to extend our records from the present. Although we cannot currently date these extremely old trees with dendrochronological methods, using carbon-14 (C14) dating techniques we can date those samples that are less than about 50,000 years old to within decades or hundreds of years of their true age and use the tree rings to potentially get more exact dates.

Expanding the search for old trees

Unfortunately, while there are quite a few bald cypress tree-ring records developed from living trees and subfossil wood that go back roughly 1,000 years, the early parts of these records typically only contain a few samples. And while the occasional discovery of unusually old bald cypress logs here and there is important and useful for particular studies, dendrochronologists who study bald cypress have long realized this is not quite enough. There is a need for a more targeted effort at locating and retrieving scores, if not hundreds, of logs old enough to extend our living bald cypress tree-ring records. In response, some of my colleagues and I began a project to work with commercial sinker-log operators to obtain samples from the logs being pulled up from river bottoms all over the southeastern United States. If you are not familiar with this process, during the heyday of commercial bald cypress logging in the late nineteenth and early to mid-twentieth century, a huge proportion of the millions of bald cypress felled in the swamps of the South were floated down rivers to the lumber mills. This was a common technique all over the United States, given the low cost of transportation with this method. Along the way, however, thousands of logs were sunk or otherwise lost.

In recent decades many smaller outfits in the timber industry have realized the enormous value of these old-growth logs—they are essentially free for the taking, with wood quality far superior to most any other wood that can be bought today. Although laws vary from state to state, typical operations are allowed to harvest these sunken logs from river bottoms or where they have floated into swamps. There is also a thriving industry of salvaging old-growth timber from structures built from the trees cut down in the last two centuries. In the southeast a great deal of these sinker logs and some of the salvage timbers are ancient bald cypress that were 500 to over 1,000 years old when they were cut. In some cases, sinker-log operators are also allowed to collect trees that died naturally and ended up preserved in the river bottom. These trees can be several thousands of years old and are particular targets of our project. Although we are only a couple of years into this project, the results have been very positive so far. With the help of several commercial sinker-log operations, we have been able to collect over one hundred samples of very old bald cypress from Florida, Georgia, and Louisiana and are working to obtain more from the Carolinas. We have very high hopes that this project will allow us to build several multi-millennial bald cypress tree-ring records in the southeastern United States and use these records to study a variety of climate change and variability issues, including droughts, hurricanes, and the large-scale atmospheric patterns that affect these and other extreme climate events.

Inspiring preservation

Perhaps more importantly than all the fascinating climate research and understanding of how extreme climate events such as drought have affected past civilizations, the knowledge that extremely ancient bald

MATTHEW THERRELL

cypress still stand throughout its range seems to inspire and move everyday people even more. Literally thousands of acres of ancient forests have been preserved because dendrochronologists shared their discoveries with the public.

Literally thousands of acres of ancient forests have been preserved because dendrochronologists shared their discoveries with the public.

Certainly nothing has done more for bald cypress preservation in modern times than Dave's discoveries of some of the oldest trees in the world. One of the best examples is the recent documentation of at least two living bald cypress trees over 2,000 years old growing within the Nature Conservancy's Three Sisters Swamp, along the Black River in North Carolina. The discovery of these trees and the likelihood of many more, some perhaps even more than 3,000 years old in the area, has resulted in an outpouring of interest and calls for additional land to be protected. In addition to finding these ancient relics, Dave and his colleagues have worked tirelessly to protect them and their habitat, even creating the Ancient Bald Cypress Consortium dedicated to protecting remaining ancient bald cypress for research, education, and conservation.

My involvement with this and other successful efforts to preserve ancient forests of all kinds by sharing our scientific discoveries has been the most rewarding aspect of my career and gives me hope that these efforts will help provide others a chance to experience the truly awe-inspiring presence of these ancient beings.

. . .

6

...

PEDUNCULATE OAK

MARTA DOMÍNGUEZ DELMÁS

SPECIES

Pedunculate oak, *Quercus robur* L.

....

LOCATION

Most of Europe up to the southern
Ural Mountains in the east

....

ESTIMATED AGE

Up to 1,000 years old

Growing up, I felt lucky enough to spend all my family vacations in a small village in the Spanish Pyrenees where my mom was born. I loved the mountains, the fauna, and the deciduous trees of those Euro-Atlantic forests. The dark green color of the tree leaves in the Pyrenees at the start of the summer vacation always made my heart beat faster. Pedunculate oaks became my favorite trees, with their beautiful lobed leaves, their acorns I would use to play kitchen, their rough bark, and strong branches to climb on. In winter, the leafless trees announced the approach of a new year, and with it, a new cycle of life: shoots sprouting where branches had been pruned, and saplings growing proudly in the now abandoned fields.

Once sacred trees for the Celts, the Germans, and other cultures across Europe, oaks have been valued as construction timber, crafting material, and fuel.

E very season, every holiday, the changing forest felt like a new playground to be discovered. My ancestors had long understood the value of these oaks—as construction timber, crafting material, and fuel. Throughout Europe, oak wood can be found in all sorts of material culture from all times. Their tannin-rich bark was used for dyeing, and the acorns provided food for animals. Some of these monumental specimens can even live up to 1,000 years, reaching 130 ft (40 m) in height and 10–13 ft (3–4 m) in diameter. It is unsurprising, therefore, that oaks became sacred trees for the Celts, the Germans, and other cultures. To me, they became the species that would determine what was to follow in my adult life.

I studied Forestry Engineering at the Polytechnic University of Valencia. There, I learned about tree growth and how trees respond to their environment. A European grant took me to Italy to study tree-climate relationships of the alpine species in the Italian Dolomites.

Gradually, I became fascinated with all the information that can be retrieved from the tree rings, and when I was offered an internship at the Ring Foundation, the Netherlands Centre for Dendrochronology, I seized the opportunity. At the Ring Foundation I worked on ecological questions about oak, beech, and pine, but one day a week, I helped with dendrochronological dating of wood from archaeological sites, historic buildings, furniture, and artworks. Most of the wood from those sites and objects was oak, centuries old but still solid and reliable. Besides historical wood, our lab also received oak samples from ancient trees that fell into the bogs and stayed buried in the mud for millennia. The first time I held a cross-section of such a bog oak I felt transported 7,000 years back in time. The building and houses around me disappeared, and I could "see" in the tree-ring pattern of that cross-section the wet environment the tree had lived in before it died. I wanted to know more. What happened to this tree? When did it die? What did these boggy environments look like and how could pedunculate oaks survive in them? The questions continued to pile up in my head, and I knew then and there that dendrochronology would be my path.

> The first time I held a cross-section of such a bog oak I felt transported 7,000 years back in time.

Oak adaptation and ring growth

About 10,000 years ago, at the start of the Holocene, oak trees began an expansive journey across Europe. As the ice retreated after the last ice age, the warming climate created more favorable conditions for forest growth, allowing oaks to migrate northward and eastward from their refugia in the Iberian Peninsula and the Balkans. This expansion occurred gradually during the early millennia of the Holocene, allowing oaks to adapt to a wide variety of environments. They found suitable habitats in diverse terrains, from mountain slopes to wetlands,

PEDUNCULATE OAK TIMELINE
• • •

Human activity in oak woodlands from
northern Spain recorded in tree rings

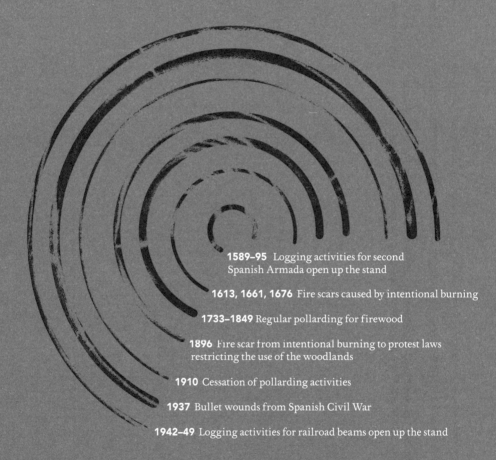

1589–95 Logging activities for second
Spanish Armada open up the stand

1613, 1661, 1676 Fire scars caused by intentional burning

1733–1849 Regular pollarding for firewood

1896 Fire scar from intentional burning to protest laws
restricting the use of the woodlands

1910 Cessation of pollarding activities

1937 Bullet wounds from Spanish Civil War

1942–49 Logging activities for railroad beams open up the stand

exhibiting distinct growth patterns and ecological adaptations depending on their environment. Oaks in Mediterranean climates, for instance, are evergreen, while those in more continental and Atlantic climates lose their leaves in winter.

Pedunculate oaks showed a preference for temperate climates, such as the low-elevation hilly landscapes of central and eastern Europe, where they became well established and often the dominant species, coexisting in mixed deciduous forests with other species like beech (*Fagus sylvatica*), chestnut (*Castanea sativa*), ash (*Fraxinus* sp.), and alder (*Alnus* sp.). In mountainous areas, they found their optimum at mid- to low altitudes, where they were less exposed to harsh alpine conditions. Therefore, pedunculate oaks cover a widespread geographical range, from the north of Portugal and Spain in the southwest of Europe; Italy and Turkey in the Mediterranean; to Ireland, Scotland, coastal Norway, and Sweden in the North; and reaching eastward into central Russia.

In these temperate climates, the growing season of pedunculate oaks ranges from April/May to September. Every spring, oaks form large earlywood vessels, through which the tree can transport water from its roots to its leaves, using energy from nutrients stored during the previous fall. Once the earlywood vessels are completed in the late spring or early summer, oaks form latewood: tissue consisting of fibers and parenchyma that supports the new biomass created by the tree every year. Latewood is formed with the nutrients available during the current growing season. Therefore, while oaks form at least one line of vessels every year, the thickness of the latewood varies according to the environmental conditions during the growing season, and in harsh years latewood can be very thin or even absent.

Ancient bog oaks

In wetland environments, including the bogs and marshes in the Netherlands and northern Germany, or in the British Isles and Ireland, oaks faced challenging conditions due to water-saturated soils. In years of high water-table, the constant moisture around the trees' roots limited their growth, and they would produce only a single line of earlywood vessels, barely enough to support survival. In drier years, the oaks would recover and once again grow healthily, producing wider rings with earlywood and latewood. Pedunculate oaks growing in wetlands thus exhibit a typical pattern of alternating periods of extreme growth reductions and subsequent growth releases, and while the species can live hundreds of years, oaks growing in bogs were vulnerable to strong winds that could uproot them when they became very large.

When these trees fell into the bog, the muddy, anaerobic environment provided ideal conditions for wood preservation, as the low oxygen levels prevented decomposition. The stems remained well preserved, buried for thousands of years. The oldest bog oaks I have ever researched were found in Roderwolde, in the northeast of the Netherlands. I could date one of them with German bog-oak chronologies, determining that it germinated around 6402 BCE and died a bit after 6135 BCE. This wood was more than 8,000 years old!

In the Netherlands, efforts to restore natural landscapes have led to the removal of ancient bog oaks from their waterlogged "graveyards." Twenty-five years after holding my first bog-oak sample as a trainee, now working as a senior dendrochronologist at Naturalis Biodiversity Center and the Cultural Heritage Agency of the Netherlands, I am holding cross-sections of ancient oaks in my hands once again. Bog oaks offer invaluable insights into past climates and ecosystems, and their removal risks erasing tangible links to our natural history, diminishing opportunities for future generations to study and connect with this environmental heritage. A balance must be found between what can be removed and what should remain buried.

Neolithic peoples and the first oak dwellings

The Neolithic period, spanning from around 5500 to 2000 BCE, marked a significant transformation in human societies, as they transitioned from nomadic groups to settled agricultural communities. Throughout Europe, a common feature to Neolithic communities settling by lake shores is the construction of their dwellings atop wooden piles. The dwellings were often built using post-and-beam construction techniques, with vertical oak posts (or piles) set into the mud to form the framework that supported the dwellings above the marshy ground. Oak is the predominant and often only species found as construction timber in many of those settlements. The choice of oak as a primary building material could have been a practical one in areas where it dominated the landscape. However, in the Early-Neolithic lake-shore settlement of La Draga (*c.* 5300 BCE), in northeast Spain, pollen diagrams show that riparian species such as laurel (*Laurus nobilis*), poplar (*Populus* sp.), and willow (*Salix* sp.) dominated the landscape around the lake, and still, 90 percent of the timbers used in the dwellings are oak. This implies a purposeful selection of this species as construction material by the Neolithic settlers.

> A balance must be found between what can be removed and what should remain buried.

Just as with bog oaks, the oak timbers of pile dwellings and crannogs have been preserved for millennia thanks to the wet and muddy environments in which they were used. Pile-dwelling sites in the Mediterranean (La Draga in northeast Spain, La Marmotta in Italy, Dispilio in Greece), those around the Alps in Central Europe (included since 2011 in UNESCO's World Heritage List), and the sites in the Baltic, offer a unique window into Neolithic life. The preserved timbers in those sites help us understand how Neolithic peoples utilized wood resources, and also reveals their construction techniques and settlement patterns. For example, we have learned that they generally selected small trees of about 8–12 in (10–30 cm) stem diameter that they could cut down with their stone axes. In La Draga, while some of those oak stems had just twenty to forty tree rings, others contained more than one hundred, indicating that the oaks, while very thin, were relatively old. This suggests the proximity of a dense natural forest with old trees, and young fast-growing ones, possibly growing at the edge.

During a second building phase, only fast-growing young trees were used. This second supply of younger oaks could have originated from acorns through natural regeneration, but also from the stumps left after the first cutting. Neolithic peoples must have then become acquainted with the capacity of oaks to resprout from the stump. Eventually, this knowledge led to a forest management practice known as coppicing. When oak trees are cut down near their base, new shoots sprout from the remaining stump, which is known as a coppice stool. Those shoots grow fast because the root system is already developed, and thus, several oak stems can be obtained in a short period of time from the same coppice stool, providing a continuous supply of small-sized timber for construction, artifacts, and fuel.

Whether coppicing was carried out as an intentional resource management practice in the Neolithic is still under debate. In the initial construction phase at La Draga, the young round trees were used for posts, while the older ones were split longitudinally for planks. This distinctive selection of trees for different purposes demonstrates that the early societies had an advanced understanding of wood properties. In this light, it is not unreasonable to think that Neolithic peoples adopted coppicing as a practice to produce a continuous supply of timber for construction and repairs, becoming the first humans consciously managing a natural resource for sustainable production. Some scholars argue that their practices irreversibly altered native forests. While this may be true, the impact of the first farmers was certainly far from what was going to come millennia later with the Roman Empire.

Oak for ships and infrastructures in the northern Roman Empire

By the turn of the Common Era (CE) the Roman Empire had expanded throughout the Mediterranean and Western Europe. With it came large-scale agriculture and husbandry, as well as mining and logging to sustain such an unprecedented economic and military power. Extensive logging for construction purposes during this period is said to have

caused deforestation in large areas of the Mediterranean, where elms, pines, and other conifer species were used for shipbuilding.

In the early years of the first century CE the Romans expanded their territory northward from the Alps along the Rhine River, constructing forts, watch towers, villages, and other infrastructure along the border as they advanced. This border became known as the *limes*. Stone was the preferred construction material in the southern provinces and the Mediterranean, but in the northern part of the border, wood, such as alder (*Alnus* sp.), ash (*Fraxinus* sp.), and especially oak, became the main building resource.

The Rhine served as a natural frontier for the *limes*, and also as a waterway to transport legions and materials from the southern to the northern provinces. The establishment of military and commercial posts in these areas required extensive infrastructure and supplies, as well as river barges—flat-bottomed long vessels purposefully designed to navigate rivers.

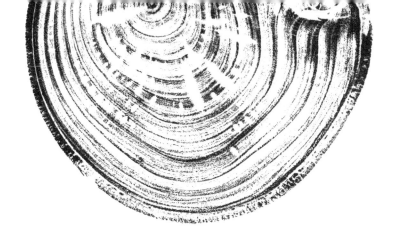

A Roman barge discovery

In 2003, during a visit to my colleagues from the Ring Foundation in the Netherlands, I joined them in the in-situ research of the river barge *De Meern 1*. When we arrived at the excavation site, the barge was almost fully exposed, and what a sight it was! With a structure of 8 ft (2.5 m) wide by 82 ft (25 m) long, entirely built with oak and almost completely intact, it was the best-preserved Roman barge ever found in the Netherlands. To find out the date of this exceptional vessel, it was decided to proceed in a minimally invasive way by cleaning up the transverse end of the timbers with razor blades, looking at the rings with a magnifying glass, and recording the tree-ring patterns directly on the timbers (as opposed to taking samples).

We estimated the width of each ring relative to the previous and the next, and gave each ring a round value corresponding to its relative width. For example, if the sequence included a narrow ring, then a wider one, an even wider one, narrow, wide, etc, we would give them values such as 50-80-100-70-100, etc, trying to maintain the coherence of relative values throughout the entire tree-ring pattern. I took turns with my colleagues lying down on planks placed on top of the Roman barge, calling out sequences of numbers that someone else would then write down on a paper sheet. Afterwards we inserted those numbers on the computer and compared the sequences of tree rings with the reference chronologies. And *voilà*! We found dates in the first half of the second century CE! I was amazed at how well this method worked.

MARTA DOMÍNGUEZ DELMÁS

Deforestation in Roman times

The scale of oak consumption for shipbuilding and infrastructure exerted significant pressure on the forests in the Netherlands. A dendro-provenance study we conducted on timbers from a Roman harbor at Voorburg-Arentsburg, in the Netherlands, provides proof of this. The trees cut around 160 CE for the first construction phase of the harbor included both locally sourced oaks from the Rhine delta and imported trees from central and southern Germany. However, the oak timbers cut in the spring of 205 CE for the second construction phase were all sourced from central Germany. Our study suggests that there was a shortage of suitable oaks in the Rhine delta by the mid-second century, and that by the early third century, local supplies of large oak timber had become completely depleted.

Overall, the use of oak for river barges, forts, and infrastructure in the northern Roman Empire exemplifies how the Romans adapted their resource strategies to local conditions, while their need to import this timber as time progressed highlights the scale of their environmental impact in local oak forests. Construction activities declined from 250 CE onward, and after the collapse of the Western Roman Empire in the fifth century CE, a period of inactivity, known as the Dark Ages, allowed woodlands to recover until population rates and construction activities rose again during the High Middle Ages.

Oak *boomstamputten* in the Late Middle Ages

Waste pits are a common feature in medieval (*c.* 600–1400 CE) archaeological sites in the Netherlands. These pits are 6½–13 ft (2–4 m) deep and dug into the sandy soil, where people threw their daily garbage. In most of these pits, several wooden barrels without bottom and lid were placed on top of each other to prevent the sandy walls from collapsing. However, in some waste pits dating to the Late Middle Ages (*c.* 1100–1400 CE) a more peculiar feature supports the walls: large hollowed-out oak tree trunks. Such tree-trunk pits (*boomstamputten* in

Dutch) have been puzzling archaeologists for decades. The size of the oak trunks (often 2½–3 ft [80–90 cm] in diameter and 13–16 ft [4–5 m] long) suggests that the living trees were even taller when they were cut down. Why use such a valuable timber resource for such a mundane purpose as waste-pit enclosure?

The choice for such large oak trees seems far from arbitrary. However, hollowing out oak tree trunks of those dimensions must have required a considerable effort: The trunks had to be split first, then each half hollowed out, to then be put together again in the pit. This process must also have generated abundant material waste, even when the woodchips could have been used as fuel. Therefore, some archaeologists propose that these *boomstamputten* may have had symbolic or ritual significance beyond their practical function, serving as votive or ritual deposition points. The discovery of valuable or unusual objects, such as ceramics and coins, within some of these pits supports this hypothesis.

However, the *boomstamputten* that I have researched offer a potential alternative interpretation. In those samples, I observed wide tree rings, as well as distorted (wobbly) growth patterns, suggesting that the trees had ample access to water and light, and likely grew on the edges of flat woodlands, pastures, or in open fields. Such trees tend to develop large, spreading branches that grow at steep angles and are vulnerable to breakage during windstorms. When this occurs, the tree is left with open wounds that are easy pathways for fungal infections. As a result,

MARTA DOMÍNGUEZ DELMÁS

the oaks can become hollow due to rot caused by fungi, making them more suitable candidates for hollowing out and less desirable for other types of timber use. This observation offers the alternative hypothesis that hollow oaks were chosen purposefully to be used as waste-pit enclosures. Instead of investing time and labor into hollowing out solid trunks, the selection of already hollow trees would have been a practical choice, allowing people to repurpose trees that maybe were about to fall, and that might otherwise have had limited use in construction.

Oaks for shipbuilding during the European expansion

The use of oak for shipbuilding has a long history in northern Europe. From prehistoric dugout canoes (the oldest ones made of oak date to around 4500–3000 BCE) and Roman barges, to Viking ships like the long *Skuldelev 5* exhibited at the Viking Museum in Roskilde, Denmark (likely of the same type used by the Vikings to arrive and settle at L'Anse aux Meadows in Newfoundland in 1021 CE), oak has been a key raw material for watercrafts and in naval architecture up until the nineteenth century. From the fourteenth century onward, oak-built ships were used by medieval merchants of the Hanseatic League for trade on the North and Baltic Seas, and EarlyModern empires such as Spain, Portugal, and England relied on oak-built ships for world exploration, trade, and warfare, making oak an essential strategic resource.

When Swedish botanist Carl Linnaeus described the species in 1753, he named it *Quercus robur*, with "robur" in Latin signifying strength, toughness, and resistance. Those qualities reflect the properties of oak, and make the wood ideal for shipbuilding. The inner heartwood is hard and resistant to decay thanks to the high levels of tannins, which also give it its dark color. Tannins inhibit fungal growth and prevent rot, conferring the preservative qualities required by a material that is going to be exposed to marine and wet environments. The outer sapwood lacks tannins, and is soft, light-colored, and prone to degradation. Therefore, it is best removed from ship timbers placed in direct contact with water.

Shipbuilders knew this. Through my investigations on shipwrecks from the Early Modern period I have learned that sapwood is indeed absent in most hull planks and other outer structural timbers from below the waterline, but that inner framing timbers often have the sapwood intact, allowing us to determine the felling date (and season!) of the trees. An exception in this regard is *Batavia*, a ship built by the Dutch East India Company (*Verenigde Oostindische Compagnie*, VOC) in Amsterdam in 1628, which sank off the coast of Western Australia on its maiden voyage to Batavia (what is now Jakarta in Indonesia). The shipwreck was recovered in the 1970s and is now on display at the Shipwreck Galleries of the Western Australian Museum in Fremantle. When I carried out the dendrochronological analysis of the *Batavia* with my colleague Aoife Daly, we found no sapwood in any of the approximately one hundred oak timbers that we sampled. Combining our results with the research by nautical archaeologist Wendy van Duivenvoorde, we concluded that VOC shipwrights had a profound understanding of oak wood properties and that such skill, together with access to a large timber-trade network, must certainly have played a role in the development of the VOC into a most successful trade company worldwide in the seventeenth century.

Shipbuilders also made conscious choices about the type of trees best suited for different structural timbers. For example, I have observed that during the Early Modern period, shipbuilders in northwestern Europe often selected fast-growing oaks for framing elements that make up the "rib" structure of ships. Fast-growing oaks have wide rings, with a large proportion of latewood. This made their wood denser, sturdier, and thus more suitable for the ship's internal framework. Fast-growing oaks (possibly managed through coppicing) could be sourced from low-elevation coastal areas of the Euro-Atlantic façade, as well as from flat and open landscapes. In contrast, hull planks of late medieval ships were made from thin radial planks. Such planks were crafted from long, straight oak stems sourced from dense forests in the north of Spain, France, the eastern Baltic, and Poland. There, the slow-growing oaks produced narrow rings with a lower proportion of latewood. The lack of large side branches in such oaks made it possible to obtain long straight planks that would be joined in an overlapping fashion, and curved to create a strong yet flexible hull that could withstand rough seas. Such a combination of slow- and fast-grown oak timber delivered an optimal balance of strength, weight, and flexibility in northern ships built up to the sixteenth century. Afterwards, designs shifted to carvel-built (smooth) hulls, while still relying on oak as the primary material until the eighteenth century.

> The oak in those artworks becomes a window into past craftmanship, historical timber trade, and its shifts . . .

The demand for oak timber for shipbuilding through the Early Modern period (*c.* 1450–1800) placed an increasing pressure on oak forests, leading to the development of policies and forest management practices in Spain, France, and Great Britain, countries that aimed to protect their oak forests and ensure a steady supply for their navies. In this way, oak was a key resource to support the expansion of trade networks and the establishment of overseas empires, playing a paramount role in shaping the course of world history.

A most valuable material for artworks

From the fifteenth to the eighteenth centuries, the Low Countries— present-day Belgium and the Netherlands—became the most eminent center of art production in northern Europe. Oak was the preferred material for sculptures, furniture, and panel paintings. For dendrochronologists such as myself, the oak in those artworks becomes a window into past craftmanship, historical timber trade, and its shifts through time due to wars or changes in woodlands.

Thousands of works of art and pieces of furniture from the Low Countries have been researched by dendrochronologists, including myself, over the past fifty years. Among many other things, we have discovered that oak from the southeastern Baltic (present-day Lithuania and Poland) was predominantly used until around 1648, when the north-European wars resulted in the redistribution of trade networks. Until then, Baltic oak was transported to trade hubs such as Amsterdam, Rotterdam, Bruges, and Antwerp, where merchants, craftspeople, and artists would acquire their wood supply.

Baltic oak was exported from Gdansk (Poland) and possibly Klaipeda (Lithuania) as wainscots, a type of timber product that results from splitting oak stems along the grain. A medieval shipwreck with such a cargo, the Copper Wreck, was found in the bay of Gdansk, carrying also a load of barrel staves and other minor timber products, besides the copper cargo that gave the shipwreck its name. The timber cargo was recovered and conserved, and is on display in the permanent exhibition of the National Maritime Museum in Gdansk. The wainscots were dated by dendrochronology to the early 1400s, when the ship was also built, providing a unique glimpse into the Baltic timber products that were sold in western markets. The high quality of this oak derived from the slow growth of oaks in the dense forests of the southeastern Baltic, which resulted in narrow growth rings and a fine grain. This made the wood particularly suitable for detailed carving, and for splitting into thin boards for panel paintings, furniture, and wall- and ceiling-paneling.

The Baltic was not the only area supplying high-quality oak wood for panel paintings. Through dendrochronology, I have identified Swedish oak with very fine tree rings in panels from the *Evangelistas* altarpiece—a splendid artwork from 1555 by Dutch artist Ferdinand Storm (1515–1556)—at Seville cathedral (Spain), and Norwegian oak in five paintings by Dutch Master Rembrandt van Rijn (1606–1669), dating between 1640 and 1644. At times of political instability, the disruption of trade through the Sound Straight, the passage between Denmark and Sweden where the Danish kings set a toll-post for ships sailing in and out of the Baltic Sea, could have been the reason why oak from western Sweden and southern Norway entered the western European art markets. Norwegian oak was available in the Netherlands for construction purposes, as our research on archaeological structures and timber frames in historic buildings has revealed. Therefore, having the same quality, and possibly a cheaper price than Baltic oak because it had not gone through the Sound toll, it is not surprising that it would also be an appreciated wood for high-quality panels.

By the mid-seventeenth century, the availability of Baltic oak in western markets was heavily disrupted by geopolitical turmoil. Artists and craftspeople in the Low Countries turned to alternative sources of timber, and the import of oak from central Germany and the Meuse River valley intensified. Progressively, wood was replaced by canvas as support for panel paintings, and by other materials such as bronze for sculptures.

Despite the shifts in the source of the wood, the use of oak for artistic production in the Low Countries prevailed until around 1750. The shifts in provenance of this material throughout that period make the story of oak used in art production not only a story of material preference, but also one of adaptation in response to historical events. The durability of oak has allowed these artworks to withstand the test of time, so they can continue to be admired in museums and collections worldwide and studied to keep on revealing fascinating stories of our recent past.

Present and future

The preference for deciduous forests that I developed as a child shaped the course of my academic choices, and my early experiences studying bog oaks sealed what would become my professional career and my passion. Most objects and structures I get to research are made of oak (possibly some of sessile oak too), and with every new sample, I feel the same excitement I did twenty-five years ago.

However, while the resilience and adaptability of oak across a diverse range of habitats in temperate climates made it flourish for millennia, the extent and vitality of pedunculate oak woodlands is far from what it used to be. The exploitation of oak forests, coupled with extensive grazing and the expansion of agricultural land during the past two millennia, has altered and decimated oak forests. Furthermore, in the past century, the preference for fast-growing species has reduced the presence of pedunculate oak within its natural range. In the north of Spain for example, where I was looking for remnants of the oak woodlands that once supplied timber for Iberian naval empires, I found extensive forest-stands of eucalyptus, planted for pulp production, where pedunculate oaks should have been.

Climate change also poses a threat to this species, as it induces insect outbreaks and other factors that cause oak decline. Modern water management practices have also transformed or eliminated natural bogs, with places like the Białowieża Forest in Poland standing as one of the last remnants of these once widespread wetland environments. As a result, contemporary oaks no longer grow in the same conditions as their ancestors, losing the unique environmental pressures that shaped their distribution over millennia. Whether this will play a role in the long-term resilience of the species, is still unknown.

Fortunately, the cultural appreciation of pedunculate oaks and oak woodlands has been growing again in recent decades. Since 2011, the yearly contest of the European Tree of the Year has had pedunculate oaks listed as candidates almost every year, with some remarkable specimens winning the competition on four occasions thanks to their exceptional stories. Such is the case of Oak Józef in Poland, winner of the 2017 edition. With an estimated age of 650 years, it is growing beside a mansion that became a cultural hub in the region, and during World War II, the tree served to hide a Jewish family from the Nazis. All these trees, winners or not, bear witness of the changes around them and connect us with the past, the present, and the future.

Conservation initiatives and sustainable forest management practices are emerging throughout Europe, aiming to protect and restore oak woodlands where possible. Organizations such as the International Council of Monuments and Sites recognize in their principles the relevance of restoring historic buildings with the same type of materials that were used in the initial construction. Such is the case of Notre-Dame in Paris. To reconstruct the nave and spire after the devastating fire that consumed the emblematic Parisian cathedral in 2019, architects looked for the same type of oak trees that were used in the original construction, finding them in their French forests.

By recognizing the value of these ecosystems not only for their wood but also for their role in biodiversity and climate resilience, we can work toward preserving and expanding oak habitats. Through responsible stewardship, future generations may still experience the richness of oak woodlands, and benefit from the ancient wisdom embedded in these remarkable trees.

· · ·

7
...
CEDRO

DANIELA GRANATO-SOUZA

SPECIES

Cedro, *Cedrela odorata*

....

LOCATION

Central and South America and the Caribbean

....

ESTIMATED AGE

Over 400 years old

Cedrela odorata, commonly known as cedro, is a tropical tree in South America well known since colonial times for its beauty, quality of wood for many purposes, and commercial value. Its aromatic, pleasant wood resembles that of the eastern red cedar in eastern North America, and it therefore became widely known as the Spanish cedar.

So far, all discovered fossils related to the *Cedrela* genus have originated in the Northern Hemisphere. Several morphological and phenological characteristics, including its leaf-shedding habit and its winged and long-lived seeds that are dispersed by the wind, indicate that cedro originates from seasonally dry forests. Evidence thus suggests that cedros migrated from the dry forests of North America to the tropical and subtropical environments of Central and South America around 30 MYA.

The presence of ancient cedro trees is not uncommon in
old-growth forests, especially in the Amazon where, if left free
of human interference, the trees can grow to be centuries old.

Later in geological times, *Cedrela odorata* also thrived in humid and wet environments and established their populations in one of the most biodiverse rainforests in the world, the Amazon. The abundance of water and energy in the Amazon rainforest allowed the trees to grow thicker and taller, but, due to its plasticity, cedro can thrive in a variety of forests. Depending on the environment in which they grow, cedro trees can vary dramatically in size from modest-sized to massive trees.

Today, cedro's territory ranges from about 26°N to 28°S and extends from Mexico southwards including countries in Central and South America and much of the Amazon Basin. Cedro thrives especially in seasonally dry tropical and subtropical forests, which are areas marked by strong seasonality and prolonged periods of droughts, consistent with the origin and physiological characteristics of the species. Cedro trees are moisture-sensitive, relying primarily on precipitation as a water source. The species' capacity to shed leaves during the dry season makes it resilient to climatic extremes such as droughts and floods. Only a few tropical forest tree species show such deciduous behavior and it is rather unique, especially considering the high biodiversity of the immense Amazon.

DANIELA GRANATO-SOUZA

Tropical dendrochronology and its challenges

Young cedro trees require light and colonize open gaps in the dense tropical forest. Despite this pioneering character early on in their life, cedro trees are a long-lived species. If the Amazon were left free from human interference, cedros could live for centuries and become large trees. The presence of ancient cedros is therefore not uncommon in old-growth forests, especially in the Amazon. I have had the chance to find and collect samples from ancient cedros, such as a 328-year-old living tree and 475-year-old cedro stump, confirming their presence in the northeastern Amazon, near the southern escarpment of the Guiana Highlands, and which have since gained special attention from scientists who aim to uncover the secrets that the forest holds.

Because cedro is one of the few tree species in the tropics that reliably forms annual tree rings that can be dated to the exact calendar year, it is one of the main tree species targeted by dendrochronologists in the region. In most places where cedro grows, the climate is seasonal, with one dry season per year. The trees shed their leaves and stop their woody and girth growth during this dry season. The resulting yearly dormancy leaves a visible mark in the wood: tree rings. Annual tree rings in cedros are easily visually identified as distinct tangential bands in the wood.

Dendrochronologists started examining cedros in the early 2000s, and have developed a growing number of tree-ring chronologies since. These chronologies have revealed many hidden secrets from various tropical regions, but cedro chronology-building can be challenging and problematic. The challenges are strongly related to the environment where the trees grow, such as soil type and competition in dense forests, and especially the climate. Areas where the amount of rainfall is constant throughout the year or where its distribution is irregular are more difficult for dendrochronology, as the trees do not experience a regular, annual dry season and dormancy and do not necessarily form annual rings. Competition with neighboring trees in a closed-canopy rainforest is also a major influence on wood anatomical anomalies, resulting in unsightly, malformed growth rings that are difficult to crossdate.

CEDRO TIMELINE
• • •

Past climate events revealed
by cedro tree rings

1695 Oldest dated cedro tree in the
eastern Amazon

1791 Drought in northern Brazil—
known to have been part of a major El Niño event

1825 Extreme drought recorded in Brazil

1859 One of the biggest floods in the Amazon's history

1865 The Forgotten Drought in Brazil

1877 The Great Famine—El Niño event affecting Brazil and India

1892 Large flood on the Negro River, central Amazon

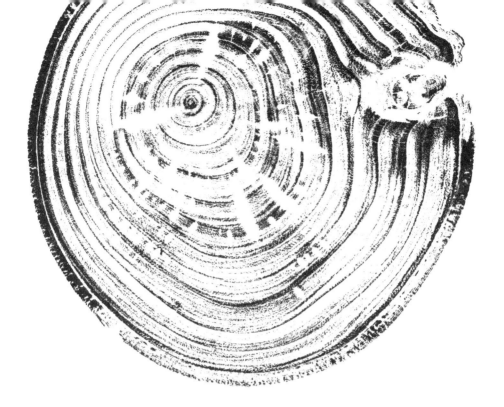

Researching cedros in the field

My first experience working with cedro was in 2016, when I started my PhD at the tree-ring laboratory in Lavras, Brazil, to develop tree-ring chronologies to study past climate in the Amazon rainforest. Ana Carolina Campos of the Federal University of Lavras and David Stahle taught me how to crossdate the trees. My PhD research gave me the opportunity to fly to remote areas of the jungle to collect cross-sections of cedro trees from legal logging operations in the eastern Amazon.

The entire process of obtaining dendrochronological samples in the Amazon was not an easy task. The Amazon rainforest is one of the most biodiverse tropical environments, which means a high biodiversity of tree species, making it difficult to identify the distribution of cedro populations. There are also logistical limitations to reaching places where there are pristine forests. Safety is another concern when venturing into the mighty Amazonian territory. The vast territory comprising the Brazilian Amazon basin is characterized by large, forested areas and floodplains that are difficult to access, where illegal activities such as logging and mining take place. In addition, the presence of law enforcement agents is very limited, if existent, which leaves researchers vulnerable and makes it unsafe for them to carry out work. Thanks to collaborations with managers of legal logging operations, we were able to obtain the first sets of samples from logged trees, a process we call salvage dendrochronology.

Being in the field has given me experience with challenges that you only learn when you are there, such as walking long distances in the forest to locate trees, determining what type of equipment is needed for the type of sample you intend to collect, and trusting the people who go into the field and guide you through the fieldwork, especially when you are a woman. I built collaborations that still exist today and worked with people who I learned to respect professionally and personally. Difficulties that I had to learn to deal with and overcome were related to the fact of being a woman trying to conduct research in a male environment. My equipment had broken down in the field because men would not listen to instructions from a woman, refused to be corrected by a woman when they misidentified a tree species in the field and tried to take away my authority in a field where I was the only female researcher, to name a few. Thanks to all the efforts and collaborators involved, we overcame the challenges.

Recently, we have started to collect samples from living trees in "undisturbed" forests that are free of human interference. While the challenges are the same—i.e. we need to find the trees, then we need to reach these remote jungle areas, collect samples, and return to our homes safely—the fact that there are no human activities (like logging operations, workers, vehicles, and machinery) in these areas adds more to the safety challenges of these field campaigns.

I recently participated in two field expeditions leading dendrochronological collections in areas where trees characterized by extreme heights have been located in the eastern Amazon. Both sites are located in the drainage basin of the Jari River, a tributary of the Amazon River.

DANIELA GRANATO-SOUZA

We camped in the jungle with no contact with civilization. We got lost for
a few hours during one of the field
campaigns, had the privilege of
sleeping at the foot of the tallest tree
in the Amazon, and collected samples
from living cedro trees. Needless to say,
the challenges were many, but they also
made for most of the fun during fieldwork and
an invaluable experience that I will carry with me for the rest of my life.

Trees growing in
stressful environments
will imprint their
difficulties on their
wood.

Crossdating challenges

When it comes to tropical dendrochronology, these challenges are just
the tip of the iceberg. Once we succeed in collecting samples and safely
returning home from our field trip, we have to face the hardest part,
which is the crossdating process. Only after we glue the samples to
wooden supports and polish their surface to get a clear view of the tree
rings, can we assess whether the trees can be dated. Unfortunately,
not all samples are datable, which can cause a lot of frustration and disappointment. Trees growing in stressful environments will imprint
their difficulties on their wood. When we think of a tree growing in
these conditions, we think of a shorter and smaller tree in girth. In the
tropics, a ten-year-old cedro tree can be the same size as a one-hundred-
year-old cedro tree. This results in all kinds of abnormal tree-ring
formations, such as false rings, very narrow and tight rings, wedging
rings, missing rings, and resin canals, fulfilling all the requirements of
wood anatomical characteristics for problematic tree-ring formation.
If most of the trees at the collected site suffer from these natural environmental conditions, the tree samples cannot be dated with standard
dendrochronological techniques.

When we look down on a triple-canopy rainforest, we see a green
ocean of beautiful trees, with the tallest trees standing out majestically,
followed by the lower-canopy trees, making up the complex vertical

structure intrinsic to these virgin forests. These trees are fighting for light, and reaching the canopy is probably their main goal in life. Although we cannot see their competing struggles, this competition leaves an imprint on the tree rings, and the suppressed growth is represented by manifold anatomical anomalies in the wood. Experience has taught me that tropical dendrochronology is very challenging and requires effort, dedication, patience, and maturity to recognize when it is worth trying and keeping on and when it is better to move on. And there are no shortcuts when it comes to dating trees accurately. I can say I have had enough experience with cedro trees to accept that I will not be able to date some of these trees.

Old cedros growing in these undisturbed tropical rainforests can live up to four hundred or five hundred years. Still, these trees are likely to show suppressed growth, which can prevent the dating and the development of a tree-ring chronology. I always say crossdating cedros is like putting together a jigsaw puzzle. Every time I come back from the field, exhausted from overcoming all the logistical challenges, I don't know if I'll be able to date the samples. It takes me time to not only analyze them, but also to find the mental strength to put together the pieces in hopes of solving the jigsaw puzzle.

When tree rings and historical records complement each other

Finding old trees and building long tree-ring chronologies are what drive us as dendrochronologists, and the reward is to have access to the stories the trees tell us. But then we have the challenge of interpreting the secrets, especially when trying to understand past climate through tree rings.

When we started analyzing the cedro cross-sections obtained from the legal logging operations in the eastern Amazon, we were surprised by the presence of well-formed tree rings and how well these trees crossdated with each other, showing a very clear common growth pattern. Wide rings are formed during wet years and narrow rings

DANIELA GRANATO-SOUZA

CEDRO

during dry years, resulting in a positive response to precipitation, meaning the cedro trees grow well during years of "normal and above average" rainfall and slow down during droughts, which would be expected given their moisture-sensitive physiology and the climate zone in which they grow.

Knowing that the width of the rings indicates how much rain fell in each year, we used the tree rings to reconstruct year-to-year changes in the amount of rainfall in the eastern Amazon. One of our wet season rainfall reconstructions dates to 1759, the other to 1786. This is more than two centuries longer than instrumental climate observations in the entire Amazon basin, which are short and often intermittent, and most stations with continuous data only date back to 1980. Such centuries-long climate records help us to better understand climate dynamical patterns such as ENSO, which is a known large-scale driver of climate extremes in the Amazon basin and which occurs on multi-year timescales. Tree-ring records that extend over centuries and thus date back prior to the start of the Anthropocene are also valuable for our understanding of the impacts of anthropogenic climate change.

The ENSO phenomenon is the abnormal warming of the waters of the eastern tropical Pacific Ocean. It is a major orchestrator of climate extremes in the Americas, leading to a dipole of dryness and wetness between the Amazon and the Pan-American mid-latitudes, resulting in an out-of-phase growth pattern between our cedro trees and the hundreds of tree-ring chronologies from mid-latitudes of North and South America archived in the International Tree-Ring Database hosted by the National Oceanic and Atmospheric Administration (NOAA). We were able to detect decadal to multidecadal variability in precipitation over the Amazon basin, which is strongly related to ENSO and to the Atlantic Ocean forcing.

Climate has always influenced human behavior, such as large migrations after prolonged droughts and the reshaping of human settlements around the world, and we can use long-lived trees to help us understand our human history alongside climate history. By studying the tree-ring chronologies we have constructed in the Amazon, we have

DANIELA GRANATO-SOUZA

been able to identify years of extreme floods and droughts imprinted on the growth of our trees, which have been confirmed by many historical records left by people living in Brazil in the nineteenth century.

The beautiful work done by historians in finding and saving ancient documents is a lifelong endeavor. It can take many years of hard work for historians to understand the past of human civilization. This is where tree-ring chronologies come into play, narrowing down specific periods of time to be investigated. The two sciences complement each other and when combined, become a rich source of information. When looking for historical documents from the eighteenth and nineteenth centuries about the history of Brazil and South America, I was very surprised by the number of weather extremes reported since the sixteenth century by Portuguese settlers and Brazilians living in the northeast and southeast of Brazil that supported the story that trees tell us.

Droughts, such as the one in 1791 that was imprinted as a narrow ring in the cedros from the northeastern Amazon, were backed up by many historical sources. Historical documents, such as newspapers, government reports, and personal letters from that time, reported on large-scale human migration, and complete devastation in northern and northeastern South America, with fields drying up, animals unable to survive, and famine setting in. Padre Joaquim José Pereira, from Rio Grande do Norte, who witnessed this extreme drought, stated that during this time, vampire bats appeared—even during the day—and attacked people and animals already starved by hunger:

> "*It was not uncommon to find houses where,*
> *like rotting corpses, poor people were still alive,*
> *lying on the floor or on the bed, covered in bats,*
> *which the victims could not even chase away.*"

Then there was the "forgotten drought" of 1865, during one of the slowest growing years for cedros in the eastern Amazon. While conducting historical research to find evidence of this drought, I found many documents talking about the "Thayer Expedition," a zoological expedition to the central Amazon in 1865, led by Louis Agassiz, the famous Swiss-American scientist who was the first to hypothesize the occurrence of ice ages in Earth's history. Agassiz and his team of scientists were joined by Brazilian scientists citing João Martins da Silva Coutinho. I found many descriptions confirming the extreme 1865 drought, including the early low water level (ebb) of the Solimões River in mid-September, two months earlier than the normal ebb in the 1860s and today. Among the many documents of this scientific expedition, there were also reports of fish kills isolated in lakes and large sand and mud banks discovered on the headlands of the islands, and even in the middle of the river.

Descriptions of the 1865 drought are reminiscent of the contemporary 2024 drought that the Amazon suffered. The early ebb of the Rio Negro, a major tributary of the Amazon River, was reported about two months earlier than normal as an unprecedented event and seen for the first time in 122 years since instrumental records began; however, it was very similar to what was reported in 1865. This made me realize how important dendrochronology is in helping us understand the natural hydroclimatic variability of the Amazon basin.

Historical records of forest fires during the great drought of 1877, one of the driest years in our tree-ring records, and the well-documented great floods of 1859 and 1892, imprinted as wet years in our tree rings, are part of the history of the Amazon, showing that perhaps today's extremes have occurred before and are part of the long-term natural variability in the Amazon basin. These ancient cedros are pointing the right way about long-term natural variability, which can be used today to help us establish what is normal and what is exceptional, and this information can be used as a basis for what actions need to be taken to avoid future disasters.

How do the cedros' ages relate to human interference in the forest?

Cedros and humans have had a close relationship since before colonial times in South America; however, as the Portuguese arrived in the sixteenth century, they started to document the timber market and expressed their preference for cedros for certain uses. Cedro was then included in the list of "royal wood" according to the Portuguese crown, which means they established a monopoly over the woods considered "royal," thus over cedros, exposing humankind's exploitative relationship with the forest since colonial times.

The wood properties of cedro make it resistant to termites, humidity, and decay, and it is easy to handle, all of which makes it valuable for furniture, interior decoration, and religious artifacts. Based on the average age of cedros among the Amazonian tree populations from which I have had the chance to collect samples and study, it is easy to identify areas where younger, fast-growing trees occur, generally toward the arc of deforestation, while centenarian trees occur in the interior, where the forests are "undisturbed." The importance of cedro for the timber industry is likely responsible for the intense, long-term exploitation and perhaps depletion of cedro trees in many regions within its native range, making it an endangered tree species that had to be protected by law in many South American countries since the early 2000s.

Restoring the connection between humans and the natural world

It is interesting to see how humans have co-evolved with the forest and its products, not just from its timber but by benefiting from fruits, seeds, medicinal extracts, and any product that does not require a tree to be cut down. Although cedro has medicinal properties and humans benefit from it, its commercial wood value appears to have had more of an impact on the depletion of Amazonian cedro populations.

Ethnobotanical studies in Brazil have shown a disconnect between humankind and the natural world, especially for younger generations, revealing how much ancient knowledge is lost. The creation of environmental laws is related to this; once people can no longer benefit from the forest, they lose interest in it. Instead of viewing the forest with affection, they see it as an untouchable resource that is in their way. Whether the protection of an endangered species under environmental law will halt its depletion and protect it from extinction is a question that remains to be answered, but the focus should be on discussions and ideas to address both of these needs: the extraction of cedro timber and its protection. Dendrochronology has many approaches that can be used to improve timber production and therefore protect the living trees from overexploitation.

Cedro in the new human-managed "natural world"

Under natural regeneration, cedro trees are scattered throughout the landscape. However, in areas with strong human interference, deforestation and selective logging (where loggers target specific tree species) determine the distribution of cedros. Sustainable logging, the legal logging activity that is currently applied in parts of the Amazon rainforest, also plays an important role in where the trees can be found. "Sustainable logging" aims for the forest to recover after a thirty-year cycle, not necessarily maintaining the same species that were there before but being able to regrow healthily. It is not known whether cedro will thrive in this new, sustainably logged landscape, which makes it a good topic for future researchers seeking to understand biodiversity in the new, human-managed "natural world."

Dendrochronology can provide information about tree growth in a way that no other science can. Knowing how fast trees grow under natural regeneration conditions is only possible through dendrochronology. By studying the tree rings and growth trajectories of trees, we understand how they behave under natural conditions in the Amazon forests. The decision on whether or not to log a specific tree, is solely based on the size and diameter of the tree. This decision-making process in sustainable logging is based on the assumption that populations of tree species can regenerate and recover trees of the same size in thirty years. Dendrochronological research has demonstrated that this is not always true, if true at all. Trees behave differently depending on the environment they grow in, and the same set of logging criteria and assumptions are not sustainable in all locations. In addition to illegal deforestation, legal logging operations can thus be a threat to cedro if the criteria applied to define cutting cycles do not consider the natural, region-specific growth patterns of the trees.

Why should we fight for our relic forests?

The main threats facing cedros are illegal deforestation, illegal logging, and, secondarily, climate change, as an unpredictable climate regime would have numerous impacts on native tree species, including cedros. When combined, these threats could severely impact the maintenance of the species in the forest. It is unclear whether anthropogenic climate change will bring dry or wet conditions to the Amazon, but it is happening at such a rapid pace that native species do not have time to adapt to the new environmental conditions, which makes them vulnerable—this is another good reason to study tree behavior with dendrochronology. *Cedrela odorata* wood is durable, light, beautiful, fragrant, and workable. And while these qualities benefit the species' ecology and adaptation to different environments, they are also characteristics that make it a desirable resource of high quality and high commercial value. In other words, its qualities come at a price. And that price is overexploitation.

Overexploitation has already caused the number of cedro trees in the Amazon to decrease. Yet despite cedro being listed as a threatened species by CITES, it is one of the species permitted for exploitation in the Amazon. This is daunting because we know very little about the complex biodiversity in subtropical and tropical forests. We don't know whether cedros continue to decline or rather thrive under the new environmental and climatic conditions. We must consider what changes are at play and what we know about cedro and its neighboring species to understand how they behave naturally and perform under change. How do cedro trees behave under high competition for light, water, and nutrients?

We know that cedro is sensitive to humidity and has deciduous habits, which can be an advantage over its evergreen neighbors in dry environments. Young cedros are light-tolerant and can colonize open clearings in virgin forests, but also human-made clearings. However, for a comprehensive picture of cedro's future, we need to consider other factors, such as their availability in seed banks, relationships between tree species that favor the success of each one in the forest, the presence or probability of invasive species, drastic changes in precipitation regimes that can harm the physiology of the tree leading to its premature death, and many other uncertainties that could alter the structure and composition of the forests in unexpected ways.

My experience in the Amazon rainforest suggests that where we find old-growth forests, we will likely find century-old cedros. Where the forest has suffered from deforestation and selective logging, we will likely only find younger cedro populations. Whether their density and spatial distribution have changed over the years is a question that remains to be answered. Science is the guide that will help us answer these questions, especially when we are able to connect different approaches to understanding the complex dynamics in all the tropical and subtropical forests where cedros now grow. The fate of the species is unclear, but if we want to preserve our ancient tropical giants, we must keep our most relic forests untouched.

DANIELA GRANATO-SOUZA

In practice, what can be done?

Cedro is an endangered tree species in many South American countries. Its CITES listing protects the species to a level that will likely do nothing against illegal deforestation or logging since these activities are beyond law enforcement's power. We, as individuals, have little power over anthropogenic climate change, as we do not have the ability to control the world's carbon emissions and all the complexity involved in this process. We can, of course, choose to live more sustainably.

So, in the short term, what would be a good strategy to protect cedro trees from being completely depleted in our tropical and subtropical forests? We can certainly provide the scientific basis for improving legal logging so that trees are not overexploited. Dendrochronology is a powerful tool to study not only the past, but also the future of this species and to gain reliable insights into the best criteria for more efficient and sustainable timber harvesting. We can and should also fight for the protection of the biodiverse forests that guard centuries-old trees.

The best strategy is to continue studying the growth and behavior of cedro trees under natural conditions. Answering questions such as: At what growth rates do they typically meet the requirements that make them eligible for exploitation? How do they interact with other species? How do they respond to changes in climate? How have they changed over the years? There are not many tree species that form reliable annual tree rings in the tropics. Cedro has proven its worth in dendrochronology, and this is a great reason why we should do everything in our power to protect these trees, especially the ancient cedros that have witnessed numerous interesting events in South America, but that are yet to be discovered.

· · ·

8

...

QILIAN JUNIPER

BAO YANG WITH CONTRIBUTIONS FROM FENG WANG

SPECIES

Qilian juniper ("祁连圆柏" in Chinese;
pronounced "*qí lián yuán bǎi*"), *Juniperus przewalskii* Kom.

....

LOCATION

Qinghai-Tibet Plateau, China

....

ESTIMATED AGE

Up to 2,868 years old

The Qilian juniper, an endemic coniferous species found on China's northeastern Qinghai-Tibet Plateau, has an extremely long lifespan, the longest-living Qilian juniper found had been fighting harsh conditions for an incredible 2,868 years, from 952 BCE to 1915 CE. Yet the tree rarely exceeds 26 ft (8 m) in height, with its largest diameter just over 3 ft (1 m). In the local dialect in northwestern China, the Qilian juniper is called "尕老汉 (*gǎ lǎo hàn*)," which means "dwarfed old man." It is small but lives long, and perhaps stands alone. Its gnarled and twisted stem and hat-like crown, give it the appearance of a petite and weathered poet, standing at the crossroads of the ancient Silk Road. With its ancient circular rings, it remains steadfast, bearing witness through time not only to incredible environmental changes but to the development of civilization itself throughout Asia.

Qilian juniper is perhaps the most important species for tree-ring studies in China, especially to study past climate, owing to its extremely long lifespan and strong climate sensitivity.

ndeed, the Qilian juniper is a symbol of resilience and longevity, not just for its wrinkled appearance but thanks to its circular tree rings, which can become as narrow as less than a quarter of a millimeter. It often grows in arid and semi-arid environments and can even survive in places with less than 8 in (250 mm) precipitation per year, such as in the Zongwulong Mountains in northwestern China. Found on southern and sunny slopes, the tree establishes itself firmly in barren, gray-brown soils with exposed rocks, often standing some feet apart from its neighbors. A juniper you occasionally see on your trips in the mountains may well have stood in the same place for a few centuries to millennia.

BAO YANG WITH FENG WANG

In Tibetan culture, the juniper's longevity forms part of a folk song, in which young locals wish their love to last as long as a juniper tree:

Above, in the realm of Tianzhu, rhododendrons bloom,

Below, in Tibetan lands, juniper stands.

Fate brought us together, a destiny entwined,

In the fourth month, we find our haven, intertwined.

If the juniper where we rest does not wither and fade,

If the seasons remain constant, forever unafraid,

Then hand in hand, we'll grow old, our love never to be swayed.

སྣེ་དང་རྐྱ་གངར་ཡུ་ལག་སྤྱི་ག་ཡུ་ག་གཅིག། སྣེ་དང་ཊ་རྫ་ང་ནགས་ཀྱི་ཤུ་གས་ང་གཉིས།།

上有天竺域杜鹃， 下有藏地圆柏树，

འགྲ་ག་དག་པ་ལ་མེ་ད་ཀྱང་ལས་འབྲང་ར་ད།། ཟླ་བཞི་བ་ཚ་མ་ད་ས་མཉམ་འད་གར་ད།།

有缘相遇是命运， 四月到来能同栖，

ཁྱ་འབབ་ལ་འ་ཡུ་གས་ང་སམ་སྨན།། ད་ས་ནམ་འཞ་ས་འགྱུར་ལ་ར་གས་བ་ནནན།།

若栖所柏树不枯萎， 若四季无变恒久远，

ལ་ཡང་ཡང་ད་གཉ་སམ་སམས་ཀྱ་ད་ར་ད།།

我俩相守共白头。

A TRADITIONAL TIBETAN LOVE SONG TRANSLATED
FROM TIBETAN TO CHINESE BY DEQIAN ZHUOMA
···········

Most natural Qilian juniper forests today are island-like, sparsely distributed across the arid and semi-arid regions of China, particularly in the provinces of Qinghai (north, east, and northeast), Gansu (south and the Hexi Corridor), and Sichuan (north). Although not the most widespread *Juniperus* species in central-western China, Qilian juniper is a dominant tree species growing on sunny mountain slopes, valleys, and ridges between 7,200 and 7,800 ft (2,200 and 4,200 m). It is frequently encountered along the ancient Silk Road from Gansu through Qinghai Province, but it does not extend into Tibet, where higher elevations make it too cold for it to survive. Qilian juniper favors mountainous habitats far away from populated towns, such as the Qilian Mountains on the northeastern edges of the Qinghai-Tibet Plateau, which is where the tree's name comes from.

Building my own route

I did my PhD work between 1997 and 2000 under the supervision of Dr. Yafeng Shi, a renowned glaciologist and Academician of the Chinese Academy of Science. Although I was supposed to work on ice cores like many other members in the group, Dr. Shi encouraged me to instead study high-resolution paleoclimate proxies and global climate change, a topic that did not yet draw strong public concern at the time. I started to shift my focus away from ice cores to the only high-resolution proxy on the Tibetan Plateau at that time, Qilian juniper tree rings. Looking back, I am still very inspired by Dr. Shi's great insight—and foresight—and I am very grateful for his open mind.

In my studies, I became fascinated by the work of Achim Bräuning, who had sampled Tibetan juniper (*Juniperus tibetica*) in the southeastern Tibetan Plateau and developed a 1,400-year-long tree-ring chronology. I decided to work in Bräuning's laboratory in Germany to learn the principles of dendrochronology. I stayed there for two years, and it was Achim Bräuning himself who brought me to the world of dendrochronology and made me a follower of Qilian juniper tree rings. After two years of hard work in Germany, I submitted my first manuscript about tree rings to a scientific journal, but the reviewers thought I had not collected enough Qilian juniper samples and unfortunately the manuscript was rejected.

I wasn't frustrated by this experience; rather, I took their suggestions in stride and realized that I must collect more Qilian juniper samples to obtain a long chronology of this tree. My team and I have since devoted many years to collecting old juniper samples, almost entirely on the northeastern Qinghai-Tibet Plateau, especially the Qilian Mountains and the Qaidam Basin in western China. We collected more than 10,000 tree cores and discs from living trees, dead trees, tomb relics from the Tuyuhun period (313–663 CE), and even wood remains buried in soils. In all this work, we overcame the difficulties of crossdating caused by the many extremely narrow and missing rings. In 2012, after decades of work and combining juniper tree-ring data from seventeen sites, our chronology covered the past 3,500 years. Four years later we extended this continuous record to 4680 BCE, making this series among the longest chronologies in the world.

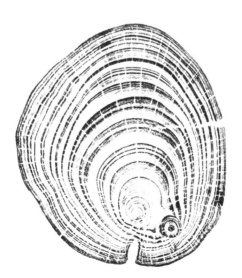

QILIAN JUNIPER TIMELINE

• • •

Key climatic epochs inferred from Qilian
juniper tree rings and historic evidence

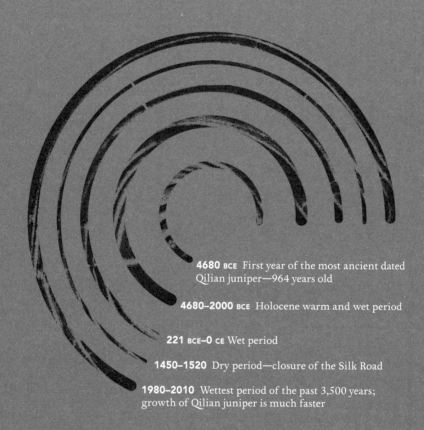

4680 BCE First year of the most ancient dated
Qilian juniper—964 years old

4680–2000 BCE Holocene warm and wet period

221 BCE–0 CE Wet period

1450–1520 Dry period—closure of the Silk Road

1980–2010 Wettest period of the past 3,500 years;
growth of Qilian juniper is much faster

Tracing back

Like other species of the genus *Juniperus*, Qilian juniper has a Eurasian origin. Its ancestors, possibly originating from the Old-World Madrean-Tethyan vegetation belts, which nowadays extend over the mid-latitudes of the Eurasian and North American continents, diverged from the *Cupressus* genus *c.* 72–50 MYA. Despite its exceptional longevity of more than 2,000 years, Qilian juniper is a recent arrival compared to its sister species. Genetic evidence suggests that it diverged from other juniper species some time during the Pliocene. Following this divergence, Qilian juniper likely underwent significant expansion, thriving across large areas as the climate and landscape permitted on the Plateau. Pollen records in lake sediments tell us this story over a broader region: During the Pliocene, forests on the Qinghai-Tibet Plateau were much more widespread than they are today, covering vast regions except for the northwest of the Plateau. During the following Pleistocene (about 2.6 MYA to 11,700 years ago), the population size of Qilian juniper remained relatively stable.

Even though no massive ice sheets were developed on the Plateau, the northeastern Qinghai-Tibet Plateau underwent a rapid transition from temperate forests to grasslands during the Last Glacial Maximum, approximately 18,000 years ago. This implies that Qilian juniper forests declined in response to this transition, forming multiple separate groups along the edges of the Plateau where great topographic variations provide microrefugia for them to survive. Qilian juniper populations near the edges of the Plateau are genetically older and those on the Plateau are the result of recolonization during the most recent postglacial period.

The northeastern Qinghai-Tibet Plateau thus has been the home of Qilian junipers for millions of years, but the juniper forests we see today do not provide a full picture of their ancient population dynamics. Qilian juniper has experienced a long and complex history marked by cycles of expansion and retreat. The environment and climate have undergone substantial changes that played significant roles in shaping these dynamics.

Transition from east to west

Today's Qilian juniper communities stretch more than 300 miles (500 km) east to west, but the areas where the trees grow are all characterized by typical alpine meadows and desert steppes. Looking up from the foot of a mountain, one would be surprised to see these relatively small trees—20–26 ft (6–8 m) high—standing as the tallest trees on rocky and sandy soils facing to the south. Even from a bird's-eye view, Qilian juniper is likely one of the few plants that can be easily distinguished on sunny slopes. Their understory features drought-tolerant herbs and shrubs, including species from the Asteraceae, Rosaceae, Ranunculaceae, and Fabaceae families.

As you travel from east to west within the distribution range of Qilian juniper, the landscapes of the forests change. A recent census found twice the species richness near its eastern distribution boundary compared to that in the west. In more favorable wetter locations in the east, such as in the northeastern Qinghai Province, Qilian juniper often forms dense forests. In these dense forests, our type of juniper is occasionally joined by Qinghai spruce (*Picea crassifolia*), a species frequently found on humid and north slopes, which forms a rare but beautiful mixed forest. In the more arid and colder regions in the west, such as Delingha, the tree more often appears in small, isolated communities surrounded by deserts, as a solitary sentinel on steep, eroded slopes. Due to the extreme aridity, dead trees can stand firmly on barren soils for thousands of years, forming an additional scenic feature in the Delingha region.

Reaching the oldest juniper

Indeed, the discovery of the oldest specimen of Qilian juniper was not an easy task. When my team and I successfully developed a 3,500-year-long ring-width chronology from Qilian juniper in western China, we told ourselves that more ancient juniper samples must be hidden somewhere. With this belief firmly in mind, we soon organized a new field campaign in the Delingha region. While a living Qilian juniper can grow for more than 2,000 years, we knew that dead trees must be the focus of this trip.

Although we decided to head to the area with numerous dead trees, we could not help enjoying the magnificent views along our way near a valley called "Zaohuogou," where endless mountains stood covered with bare rocks and patches of juniper under a clear sky. As no real paths were accessible in this remote area, we spent nearly five hours climbing about 1,600 ft (nearly 500 m), from an initial elevation of about 12,000 ft (3,650 m) to the top of a mountain at 13,60 ft (4,150 m), using the trails typically walked by blue sheep. Even if dead trees were our targets, sometimes we still stopped to collect cores from living trees that looked potentially old, making our way toward the ridge slower.

Compared with living trees, sampling dead trees was more challenging. We tried to collect 4-in (10-cm) thick disks from the trunk of dead trees if they were not completely standing. As we sampled more, the number and weight of the discs increased. After a long day of fruitful and intensive fieldwork, we had collected a total of thirty disks. On our way back down, I noticed a stump of a dead tree that was hidden almost entirely under the soil. Everyone thought we could give up on this tree, since only a couple inches of woody tissue were exposed on the surface, but I insisted on digging it out of the soil. Subsequent crossdating back in the lab proved this sample was the oldest we've ever collected, dating between 4680 and 3715 BCE. If I had not insisted, we might have missed this precious sample dating over 6,000 years old!

A thriving survivor

Qilian juniper was "born" to adapt well to high-altitude and mountainous environments, as thriving in a harsh habitat seems to be the overriding theme in its life. Its distribution range is characterized by a continental arid climate with cold and dry winters and relatively cool and rainy summers. Annual temperatures average 23–37°F (–5 to 3°C) and the annual rainfall ranges from 10 to 21 in (370–530 mm).

This means both heat and water supply become critical challenges that Qilian juniper must overcome—which it does—in its own unique way.

Qilian juniper is smart. It chooses to grow on south slopes where sunshine is abundant. Its seeds are stored in turbinate cones, which are relatively large and heavy, thus limiting the juniper's chances to climb to higher, harsher elevations. The seeds are also adapted to germinate in nutrient-poor soils, ensuring the tree's persistence.

Qilian juniper leaves present in needle-like and scale-like forms and are coated with a waxy layer, which protects them from drying winds and intense sun radiation at high altitudes. The needles also have a thick cuticle and sunken stomata that minimize water loss and reduce the risk of frost damage. Moreover, the proportion of scale-like leaves increases with tree age. This reflects a strategic transition during different stages of a juniper's life—needles conserve more water during the juvenile period, and scale-like leaves with larger surface areas are more efficient at photosynthesis in a mature tree that requires more "food" to grow. In addition, the evergreen leaves of Qilian juniper can photosynthesize at a relatively low temperature to best take advantage of the sunshine during a short growing season.

Zooming out, Qilian juniper ends up making correct decisions on macro scales to adapt to its harsh environment. Tree crowns are typically narrow and irregular, looking like a hat, and branches and the stem align spirally upward. To form a more solid base, the tree grows its roots deep into the ground. These elegant yet sturdy tree structures help the tree to survive in strong mountain winds.

Grow slow but stay long

Maintaining slow growth is one of the Qilian juniper's most fascinating strategies, and one that has allowed it to potentially live for thousands of years. From a cohort of twenty trees all exceeding two thousand years in age, we found that the Qilian juniper's stems on average grew by less than a quarter of a millimeter in diameter per year. The narrowest measured ring only consisted of two layers of cells; one formed in summer

and the other in early fall. In contrast to common knowledge that trees form a new ring each and every year, in extremely dry years a juniper tree may not even produce a ring at all. Dendrochronologists call this phenomenon "missing rings." The rate of missing rings is as high as 5 percent in Qilian juniper, suggesting that this tree has sufficient flexibility to allocate little energy for stem growth in "bad" years. As a result, Qilian juniper trunks rarely exceed 3 ft (1 m) in diameter, even for the oldest individuals more than a thousand years old.

We can see a similar strategy in terms of the tree's height. In exposed locations, Qilian juniper can reach 33 to 66 ft (10–20 m) in height, but old trees frequently remain stunted and closer to the ground, standing only 16½ to 26 ft (5–8 m) tall. This limited tree size and slow growth rate also make the wood structure denser, all of it providing strong structural support to help Qilian juniper better resist damages from strong wind as well as snow and ice accumulation.

Chinese ancestors and Qilian juniper

When the first Chinese settlers arrived on the Qinghai-Tibet Plateau, their life was a constant struggle against poor natural resources and harsh climatic conditions. Qilian juniper proved to be an excellent construction material and a durable fuel, as the tree is widely distributed, slow-growing, and its wood is of a high quality. Even today, we can find building supports made from Qilian juniper trunks in excavated archaeological sites, such as in the tombs of the Tuyuhun Tribe (313–663 CE). In fact, archaeologists and botanists have successfully deduced the age of the Tuyuhun Tribe's existence using the wood of Qilian juniper in those tombs. These well-preserved wood remnants are still offering opportunities for tree-ring research in deciphering climate change.

Our ancestors also discovered the medical value of Qilian juniper. They believed that the smoke from burning juniper branches possesses the power to spiritually purify "evils," a practice that can be traced back to the "Wui Sang" (Tibetan: ས་ དབངས་, or *bsang*) prayer rituals of the ancient Xiang Xiong period, around 4,000 years ago. Although this ritual is mystical and religious, it is closely related to the unique medicinal value of the Qilian juniper. The ceremony has been widely integrated into various aspects of Tibetan daily life, such as weddings, funerals, send-offs, and homecomings, and continues to be in use. In fact, some of these beliefs about Qilian juniper's medical functionalities have been incorporated in traditional Chinese Medicine. For example, its cone and small branches are considered effective hemostatic agents.

The start of the tree's circle

Qilian juniper is perhaps the most important species for tree-ring studies in China, especially to study past climate, due to its extremely long lifespan and strong climate sensitivity. As early as 1975, Zhengda Zhuo in Lanzhou University led one of the earliest modern tree-ring studies in China on Qilian juniper's tree rings in Qinghai Province. Zhuo was able to build a 917-year long Qilian juniper tree-ring record, even

though crossdating had not yet been
introduced to Chinese researchers at
the time. His climatic interpretation
of the tree-ring record has been confirmed in the scientific literature
nearly half a century later, after crossdating became a widely used technique
among Chinese dendrochronologists.

*... in extremely
dry years a juniper
tree may not even
produce a ring
at all.*

Since the 1990s, Chinese geographers and
botanists have developed great enthusiasm to search for the oldest Qilian
juniper trees, in a quest to build the longest-possible tree-ring and paleoclimate record in China. In 1997, Xingcheng Kang developed the first
crossdated Qilian juniper ring-width chronology surpassing 1,800 years
from Dulan County. This record was soon extended to 2,326 years by
Qibing Zhang in 2003, 2,485 years by Yu Liu in 2009, and then to 3,585
years by Xuemei Shao for the Delingha region in 2009. These long
records have greatly improved the understanding of past year-to-year
changes in western China's climate at annual resolution. Finding more
ancient Qilian juniper trees is therefore the dream of every Chinese
tree-ring researcher, and it is mine as well.

Geochemical insights

As dendrochronology develops, new techniques open additional ave-
nues for extracting climatic information from tree rings. A recently
popular method to develop proxies of past climate is the use of stable
isotopic ratios of tree-ring cellulose, measured through mass spectrom-
etry. One of the advantages of such tree-ring stable isotopes is that you
do not need as many samples as you do for ring width to represent past
climate reliably and accurately.

Measuring stable isotopes is much more labor and cost intensive
than old-fashioned tree-ring width measurements, but I was motivated
to try this technology to decode climate information from the most
ancient Qilian juniper trees that date as far back as the Middle Holocene.

For this ancient period, more than 6,000 years ago, only few samples are available, but stable isotope measurements have the potential to overcome the problem of limited sample availability. For more than five years, our group has tirelessly undertaken the monumental task of slicing woodchips, extracting cellulose, and measuring isotope ratios on thousands of tree rings.

All this hard work has finally paid off and resulted in a 6,700-year-long oxygen stable isotope chronology, which not only reveals precipitation variability in the Delingha region, but also contains information about the East Asian Monsoon because its boundary crosses the distribution area of Qilian juniper. The East Asian Monsoon brings a substantial amount of moisture from the Pacific Ocean and Indian Ocean to East Asia during summer. It is critical for China, Japan, the Korean Peninsula, and Southeast Asian countries. This series was iconic for many reasons, not least because it shows changes in the Monsoon on millennial and multi-millennial timescales, which is not possible with shorter records. However, our new record agrees very well with records derived from other archives such as sediments, loess, and stalagmites in monsoonal China, a region where many ancient Chinese civilizations flourished.

The East Asian Monsoon
and ancient Chinese cultures

Using our new East Asian Monsoon reconstruction, we were able to further understand the relationship between monsoonal climate change and social resiliency in China from a longer-term historical perspective. We found that the period from 4680 to 2000 BCE, called the Holocene optimum, was 40 percent wetter than the average over the entire Holocene period. This relatively humid climate, caused by a northern position of the East Asian Monsoon, was beneficial for and facilitated the spread of agriculture, allowing millet and millet by-products to flourish. The Holocene optimum corresponds well in time to the period of the Yangshao culture in the Yellow River Valley (*c.* 5000 BCE to 3000 BCE), which had mastered painted pottery techniques. The Majiayao (*c.* 3300 BCE to 2000 BCE) and the Qijia (*c.* 2200 BCE to 1600 BCE) cultures, scattered in the upper reaches of the Yellow River in Gansu and Qinghai Provinces, also thrived in the late phases of this humid and possibly warm Holocene optimum. The earliest documented Chinese dynasty, the Xia Dynasty (*c.* 2070 BCE to1600 BCE), similarly emerged near today's Henan Province during this period.

From 2000 BCE onward, the climate changed from humid to arid, and culminated in a drought around 1600 BCE. This dry period was possibly related to the three-way split of the Qijia Culture, and is believed to have resulted in the collapse of the Xia Dynasty. Indeed, the droughts might have accelerated the disintegration of historical civilizations spanning the early Neolithic to early Iron Age (*c.* 8000 BCE to 500 BCE), and at the same time led to a precipitous decline in the number, as well as the distribution area, of archaeological sites throughout China.

The interaction between monsoon changes and societal changes is more rigorously illustrated from a few centuries before the Common Era onwards, when Chinese history is better documented. When a strengthening of the monsoon from circa 1000 BCE to the start of the Common Era resulted in another wet regime, prosperous cultures reoccurred over most of China. A symbolic societal transition that occurred during this humid period was the unification of most of the country by the Qin Dynasty in 221 BCE. Although the Qin Empire disintegrated rapidly, the Han Dynasty soon thrived and tremendously influenced many Eurasian civilizations.

Juniper witnesses the Silk Road

The Silk Road, connecting ancient China with the Western world from the second century BCE to the mid-fifteenth century, was extremely important in Chinese history. Consisting of a network of multiple trade routes, the Silk Road starts from Xi'an (ancient name Chang'an), the capital of the Han Dynasty and stretches westward along the Hexi

Corridor and the Tarim Basin, to western Asia. The Silk Road is no longer functioning today, but its prosperity and influence is clear from historical remnants, including massive Byzantine coins excavated along ancient routes. Marco Polo, the well-known Italian explorer and the author of the *Book of the Marvels of the World*, traveled to many ancient Chinese cities via the Silk Road in the thirteenth century.

Qilian juniper trees witnessed the birth and decline of the Silk Road. Our isotope-based reconstruction indicates that the relatively wet climate during the Han Dynasty (Western Han from 202 BCE to 9 CE) provided ideal conditions for producing abundant food and water sources that supported the ancient Chinese armies to expand westward into today's arid regions. The Han general Huo Qubing was known for defeating the Xiongnu (also known as the Hun) Tribes during this period. Consequently, Zhang Qian, the first official envoy of the Han Dynasty to the West, opened the Silk Road along the Hexi Corridor, influencing many Asian civilizations through commercial trades but also facilitating cultural, technological, and religious exchanges. At the same time, we also noticed in our reconstruction from Qilian juniper tree rings an exceptionally dry period from the 1450s to the 1520s that coincides with the closing down of the ancient Silk Road. Although few historical documents have recorded the exact reasons of this societal shift away from the Silk Road, Qilian juniper offers strong evidence of how climate change could have had an influence on human society in this case.

The circle continues

With origins dating back millions of years, the Qilian juniper has evolved miraculous survival strategies—its evergreen needle-like leaves, semi-fleshy, flexible branches, and deep root systems all improve its chances to thrive in the cold and drought of the Qinghai-Tibet Plateau. However, present-day Qilian juniper forests are facing formidable challenges from anthropogenic climate change and intensifying human activities. Rising temperatures have undoubtedly heightened the tree's evapotranspiration stress and water demand, particularly for

QILIAN JUNIPER

lower-latitude and older trees that are more vulnerable to drought. Even though western China has become wetter over the twentieth century, precipitation has become increasingly erratic and unpredictable, signaling greater risks of future weather extremes. The eastern Qilian Mountains in the northeast of the Qinghai-Tibet Plateau, for instance, suffered exceptionally low rainfall in the summer of 2023, which starkly contrasts the general humidification trend in the region. Such extreme droughts threaten the habitats of Qilian juniper, leading to widespread mortality, but also increased precipitation presents a double-edged sword for the Qilian juniper communities.

A wetter climate may provide more water but can also intensify ecological competition as other vegetation thrives, can accelerate the decomposition of ancient junipers that have stood for millennia, and trigger massive insect outbreaks. In addition to this, the expansion of herding continues to impact the resilience of the Qilian juniper ecosystem and can alter its population dynamics. Projections indicate that northwestern China will continue to warm and become more humid through at least the end of the twenty-first century. These future climate changes alone can drive dramatic declines in the population of *Juniperus* species, including the resilient Qilian juniper.

Given the profound role of Qilian juniper in its ecosystem and its remarkable capacity for long-term carbon sequestration across diverse mountainous landscapes, the future of this tree and its surrounding habitats demands our deepest attention. Thankfully, in recent years, concerted efforts have been made to safeguard both the ancient and young Qilian juniper populations. Among these endeavors is the establishment of the Qilian Mountain National Park dedicated to preserving the rich resources of wildlife and vegetation on the northeastern Qinghai-Tibet Plateau. At the same time, policymakers are actively engaged in managing and cultivating artificial Qilian juniper forests, driven by a desire to conserve soil and water, promote agroforestry, and enhance the landscape's natural beauty. These initiatives reflect a growing commitment to protect these resilient sentinels of time and the fragile ecosystems they inhabit.

Retaining the circle of life

Just as the Qilian juniper quietly and steadfastly records the history of environmental change in its habitat, scientists such as myself continue to tirelessly unravel the climate and life codes inscribed in every ring of this tree's long life. We seek to understand the ways in which the environment has transformed over time and how these ancient trees have endured such conditions, aiming to better comprehend climate variability and safeguard the species in a world of change. In the Qilian juniper, we have discovered a fascinating story of resilience and shifting landscapes, offering deep insights to myself and all those who study it. When humanity finally learns how to protect this remarkable species, it will likely be the Qilian juniper itself that has taught us how to ensure its circle of life remains unbroken.

Regardless of whether we learn to protect the Qilian juniper, this oldest and most enduring conifer in western China will remain—whether in life or death—interwoven with the ecosystems and human societies around it. As a vital natural archive, its enduring story will persist, forever unveiling the chronicles of climate and environmental change across the vast expanse of time. And while its ancestors have faded into history, the circle of life for this ancient tree does not need to come to an end. It has an incredible ability to endure, this poet at the crossroads. Further understanding will allow it to continue to bear witness, as it has done for thousands of years, recording tales of changing climate, evolving landscapes, and human civilizations, hopefully long into the future.

• • •

9

...

GREAT BASIN BRISTLECONE PINE

MALCOLM K. HUGHES

SPECIES

Great Basin Bristlecone Pine, *Pinus longaeva*

....

LOCATION

California, Nevada, Utah, USA

....

ESTIMATED AGE

5,000 years old

You could see the Great Basin bristlecone pine as a very unlikely "ancient tree." It lacks the majestic stature of giant sequoia in California, kauri in New Zealand, or alerce in Chile. It does not provide the encompassing green embrace of an ancient oakwood in Europe. Instead, these ancient trees are scattered thinly across rocky slopes, each tree with its unique, scrubby form, but a few feet tall, that looks only just alive. Yet Great Basin bristlecone pines have been hanging on, near the edge of survivability, for millennia. Their most striking attribute is persistence. Thanks to this, Great Basin bristlecone pines are the world's oldest living trees.

*Living in harsh environments where few other plants thrive,
the Great Basin bristlecones are ancient survivors that are helping
us understand the events and processes that shape our world.*

One, high in the Great Basin National Park in eastern Nevada, was felled in 1964. Its innermost ring, a few feet above the ground surface, grew in 2936 BCE. Had it survived to 2024, its age would have been 4,960 years. This species grows slowly in early life and the tree probably took a few decades to reach the height sampled. So, 5,000 years is a cautious estimate of its longevity. Two trees of this species in the White Mountains of California are known to come close, with ages greater than 4,800 years. Many hundreds of other individuals are known to have survived millennia in the high, cold, dry mountains of California, Nevada, and Utah. Scrappy, tough westerners! They provide a treasure trove of data, questions, and inspiration for generations of researchers, past, present, and future. As a young ecology lecturer in early 1970s England, I started my journey among the green, lichen-clad, soggy oakwoods of the western side of the British Isles. Studying these forest ecosystems inevitably led me toward events on timescales of decades and centuries. For this, it became clear that a reliable timekeeper was needed. That's where tree rings came in.

MALCOLM K. HUGHES

Against the expectations of colleagues with far better "letterhead" than I, it turned out that my up-to-250-year-old sessile oak (*Quercus petraea*) near sea level in north Wales had strong common patterns in the width of their annual rings. This meant that each annual ring could be robustly assigned a calendar year. So did living and bog oaks in nearby Ireland and Scotland, being studied by colleagues in Belfast to build back thousands of years to check the then quite new radiocarbon timescale. About 4,970 miles (8,000 km) west, long-lived Great Basin bristlecone pines identified in the early 1950s were also proving useful to tree-ring dating. Despite their distance from the oaks, these scattered stands of bristlecone—living trees and their older remnants—found in the high, dry mountains of eastern California and the Great Basin, had been recruited to check the radiocarbon timescale as well. So it wasn't long before my quest to understand and exploit tree-ring dating took me to work at the University of Arizona's LTRR. This in turn led me to the exquisitely beautiful but bracing landscapes where the world's most long-lived trees—the bristlecones—grow.

The subgroup of pines that includes Great Basin bristlecone pine (*Pinus longaeva*) includes two other species found in the western United States: Foxtail pine (*P. balfouriana*), found in California and Oregon, and Rocky Mountain bristlecone pine (*P. aristata*) found in Arizona, Colorado, and New Mexico. In this chapter, a reference to "bristlecone" means "Great Basin bristlecone pine."

Nature's timekeeper

Great Basin bristlecone pines provide a treasure house of year-by-year information on an amazing range of aspects of life on Earth. Relict wood remaining on the open rocky ground of the Great Basin's high mountains was left by trees that grew just after the planet emerged from the most recent ice age, more than 12,000 years ago. We can decode a wealth of detailed information by combining the annual rings of old living trees overlapping those in relict wood from the same places. This includes far-reaching scientific advances such as the calibration of the radiocarbon timescale as well as records of the effects of distant volcanoes on global

GREAT BASIN BRISTLECONE PINE TIMELINE

• • •

Climatic and Earth system events
evidenced by Great Basin bristlecone pine

9294–9243 BCE Likely earliest ring, Methuselah Walk (MWK)

6827 BCE First year, dated chronology at MWK

400–1500 CE Severe droughts shown in lower-elevation rings

536 CE Frost rings, blue rings at upper tree line, volcanic dust veil

774–5 CE C14 spike at MWK—major solar storm

1328 Lowest tree line in White Mountains—persists to 20th century

1458 Frost rings at upper tree line, low-latitude eruptions

1950–onward Fastest ring growth near tree line since 1750 BCE

climate. They also reveal centuries-long droughts in western North America, track the decline of the upper tree line over the millennia, and help show the unique character of the climate of the past few decades, both globally and locally.

Persistence is key

How does bristlecone wood last so long, in living trees and dead logs? Amazingly, there is no evidence for biological senescence—aging—in Great Basin bristlecone pines. In fact, as Ronald Lanner states in *The Bristlecone Book*, they show an "absence of an aging effect in the shoot apexes and cambium." He further notes that their ability to reproduce successfully doesn't seem to be impaired, even at an advanced age.

Several factors combine to leave bristlecone wood intact both before and after tree death. It is extremely dense, growing very slowly. It is also extremely resinous and, in its cold and dry environment, resistant to fungal decay. Scouring by wind-borne dust and ice crystals weathers dead bristlecone wood where it stands or has fallen, evoking a comparison with driftwood on a shore. Yet it can remain completely sound internally for ten millennia or more. This is helped by the fact that conditions are so limiting for plant growth in the places where these oldest trees and wood relicts remain that there is little other plant life—thus little fuel capable of carrying wood-consuming fire. Even with these unique preserving traits and conditions, how do these trees survive the repeated physical assault associated with their typically high, dry, rocky, windy locations?

Key to this tree's survival is its ability to limit damage. Although each tree was likely derived from a single seed, it responds as if it were a bundle of independent connections between roots and needles. This is called "sectored architecture." Imagine a major root is broken by erosion of the slope on which the tree grows, often on its downhill side. The resulting lack of water will stop wood formation and all kinds of shoot, needle, and reproductive growth right up to the top of the tree in that sector. The cambium dies and with it the bark. The other, unaffected, sectors continue to grow as before, with new tree rings, green needles, root, and shoot growth.

As the hill slope erodes or other disruptions occur, other sectors die, leaving more and more wood exposed. A narrower and narrower band of bark (and so of phloem and cambium) persists, under which the exchange of water, nutrients, and hormones continues to support wood growth and the canopy of the connected branches above. This form of growth is called "strip bark." The annual wood growth no longer occurs in complete rings around the stem, but as bands under the surviving bark. A single strip of bark may cover a continuous series of thousands of such bands. Without sectored architecture, the original break of a major root could lead to the death of the whole tree. With it, the damage is limited to the injured sector, allowing part of the tree to survive.

This dance of trauma and isolated response, combined with the way living branches develop to accommodate and correct for leaning, further helps the tree persevere. It also results in the strange, complex, and individual forms of these trees as they age. Even more irregularity is produced by buds within the needle bundles that, after years of growth around them, can produce new branches, facilitating recovery of the tree's canopy after damage.

Conifer survival

This ancient tree's ability to survive is also due to its conifer origins. Part of the legacy of being a conifer is a greater ability than in most flowering plants to foster certain soil fungi called mycorrhizae. These aid in the extraction of water and vital nutrients from poor soils. Another is the lack of big leaves replaced every year that capture a lot of light but lose lots of water vapor to the atmosphere. Such leaves are often attractive and nutritious to insects and other herbivores. In their place, conifers have slim, unpalatable needles. A bristlecone's needles are kept for decades, buffering the effects of year-to-year weather fluctuations by adjusting needle length—strong growth yields a whorl of longer needles; a poor year, shorter needles— and by storing in them carbohydrates in reusable form for hard times.

For many conifers, these hard times can also include insect attacks that lead to increased mortality. In recent decades, numerous pine populations

in western North America have suffered from mountain pine beetle (*Dendroctonus ponderosae*) or pinyon ips (*Ips confusus*) outbreaks. The initiation of these outbreaks can be associated with conditions stressing tree growth, especially hot droughts. Bristlecone and foxtail pine are less susceptible to these attacks than other pines. Not only are they better adapted to resist water stress but the chemical defenses provided by their resin are particularly effective. So far, the cool temperatures at high elevation have restricted mountain pine beetle to a single generation of larvae per year, although that may change in a warming climate. Even so, it is vital that the bristlecone stands be carefully monitored and protected from fire risk, as well as for the nearby spread of the beetles and their other hosts.

This ability to recover as well as to survive hard times—effectively "riding over" the good and bad times—means that bristlecones don't have to recruit successfully all the time. Their seed germination is also helped by the Clark's nutcracker bird, which shallowly stores some of its foraged bristlecone seeds in any soil it can find. Uneaten seeds then have a chance to germinate and start a multi-millennial life, taking their time to see through any climactic "bumps" in the road.

In addition to its extraordinary ability to persist, the bristlecone pine tends to reliably form one ring a year over the surface of its trunk, branches, and roots. This makes it particularly suited for tree-ring dating. Distorted rings only form quite rarely—for example, they may form from late spring or early fall frosts that damage still-forming tracheids (the cells forming the tubes—the plumbing—reaching from roots to needles). Only very rarely is no ring formed at all, and this can, in fact, prove helpful to see wider implications across a large area. For example, the year 1580 was so bad for ring growth that many trees across the southwestern quadrant of what is now the United States formed no ring, with drought the likely cause.

Discovering—and expanding on—the "living ruins"

The potential of bristlecones for tree-ring research was first realized by Edmund Schulman (1908–1958). He traveled by car, on foot, and horseback to many remote mountain regions, searching for suitable trees to develop yearly maps of climate in semi-arid regions. Such trees would be old, to extend the climate record as far back as possible, and their ring growth would vary strongly from one year to the next, implying year-to-year climate variability.

The MWK chronology is now the world's longest replicated tree-ring chronology from one tree species at a single, small location.

Finding the long-lived bristlecones in 1953 was a by-product of this visionary project. In *Edmund Schulman and the "Living Ruins"* Donald J. McGraw describes Schulman's "side trip" to the White Mountains near Bishop, California in 1953. This was when the "first known tree over 4,000 years of age was identified." Over the next few years Schulman found and crossdated seventeen trees over 4,000 years old there, including one that lived more than 4,600 years. He found the oldest trees in what is now called Schulman Grove, near the lower elevational limit of the species in those mountains. Only three months after his death, in 1958, *National Geographic* published his article, "Bristlecone, Oldest Living Thing."

Schulman's former assistant, Charles Wesley Ferguson (1922–1986), took this further. Through the 1960s he sampled and crossdated many standing snags (dead trees still standing) and fallen logs, mainly from Methuselah Walk (MWK), an area of about 250 acres (1 sq km) at around 10,000 ft (2,800 m) above sea level within the Schulman Grove. By 1969 he had extended this site's chronology back 7,484 years (since extended to 8,837 years' length). In 2013, Matthew Salzer and I calculated that of 493 measured MWK samples, more than 125 had more than 1,000 rings, and 21 had more than 2,000. Every year back to 6366 BCE was covered by at least ten samples!

The MWK chronology is now the world's longest replicated tree-ring chronology from one tree species at a single, small location. There is even older wood at MWK, pieces whose ring patterns match one another but not the crossdated almost-9,000-year-long site chronology. Recent work on these "floaters" has raised the prospect of extending the MWK chronology back as far as 9294 to 9243 BCE. Ferguson's efforts had taken the wonderful gift of Schulman's discovery and more than doubled its length.

Using the network Schulman started

Over the next four decades, dendrochronologists continued building on Schulman's network of climate-sensitive long-lived conifers in western North America, collecting and analyzing thousands of samples, finding many trees and remnants from Schulman and Ferguson's original samples. The heart of this network was Great Basin bristlecone pine and in turn the MWK site chronology from the lower, drier elevations at which the species grows. We showed that this chronology contained good information on the year-by-year variability in annual precipitation for the past 8,000 years over much of the Great Basin. The recent end of the record captured known droughts, for example, the single-year 1976 drought, which was followed by a switch to wetter conditions and then the multi-year drought of the late 1980s. Many such droughts and some more severe were recorded over the eight millennia. So, at least up to the 1990s, such droughts were not unusual.

A smaller, tantalizing, portion of the variability was slower, made up of wetter and drier periods of decades or even centuries. Could they be real? Might they be phony—products of chance or of our methods? To check this, we used a network of six bristlecone site chronologies, all from near the lower, drier limit of the species. It included MWK in California and reached to Utah. The group of site chronologies shared the same precipitation signal as MWK. For most of the past 2,000 years they showed a common decades-to-centuries pattern. There were two very sustained droughts from the tenth to the fourteenth centuries CE. These droughts also show up in other tree-ring records from the Sierra Nevada and, dramatically, in low stands of nearby Mono Lake, a closed basin that lost water only by evaporation. These historic Mono Lake low stands occur in a completely independent record of paleo beaches.

We can use tree rings to reconstruct the flow of water into Mono Lake and put this flow into a computer model of the shape of the Mono Lake basin. As the model lake filled up in times of high inflow, the level of the water and hence the beach level rose. As the model lake level fell, in dry times, so the carbonate chemistry-encrusted remains of life in and around the lake emerged. When using these remains and radiocarbon dating to

plot the lake water level over centuries, the history of the lake level variations shows low stands of the lake uncannily close in time and actual level to those generated by our tree-ring driven model. This is evidence that the multi-decade to century-long regional droughts indicated by the bristlecone were real and part of our climate well before the Industrial Revolution.

Tree lines and climate change

In 1973, Val LaMarche used remnant bristlecone wood to estimate the history of the bristlecone pine upper tree line in the White Mountains of California. He used this to infer climatic change over the millennia, especially summer temperatures that could set the upper elevational limit of tree establishment and survival. He found wood far above the modern tree line from as far back as 5400 BCE. The highest tree line firmly evidenced by remnant wood was at least 490-ft (150-m) higher than the recent level.

From this, LaMarche inferred July temperatures 3.5°F (1.94°C) warmer than in the early twentieth century. The tree line remained at about this level from 4500 BCE until a sudden decline of about 330 ft (100 m) around 2200 BCE. It was followed by a more gradual, episodic decline until reaching its lowest levels, and lowest inferred summer temperatures, by 1100–1550. In 2013, our group, led by Matthew Salzer, built on LaMarche's work by a great deal of new sampling and tree-ring analysis from both Sheep Mountain in the White Mountains, California, and Mt. Washington in the Snake Range, Nevada, more than 230 miles (375 km) to the northeast. The resulting broad pattern was similar to LaMarche's findings from the Whites. The regional similarities in the long-term patterns of tree line change in California and Nevada pointed to a mainly temperature-driven cause. As LaMarche reported, we, too, found young Great Basin bristlecone pine up to 240 ft (73 m) above the modern tree line. We stated that, "It is likely that full-size trees have not grown at this elevation since before 2200 BCE and quite possible that high elevation ecosystems are responding in a manner unprecedented in approximately 4,200 years."

As LaMarche and others were filling out Schulman's dreamed-of network, they became aware of an acceleration of ring growth since the mid-nineteenth century. This happened in high-elevation bristlecone and some other tree species in the same region. They could see no climate cause when comparing to the weather records then available for these high mountains. They suggested that, since they could see no climate cause, direct fertilization of tree growth might be at work. They proposed this fertilizer was the increasing atmospheric concentration of carbon dioxide from fossil fuel burning. Supplemental carbon dioxide is, after all, sometimes used in greenhouses.

Again, we re-examined LaMarche's suggestion. We had the benefit of more and better climate data than that used twenty-five years earlier. We also used a set of closely spaced site chronologies along an elevational transect in the White Mountains and tree-line site chronologies from hundreds of miles distant in Nevada. We saw that, near the tree line in all cases, "ring growth in the second half of the twentieth century was greater than during any other 50-year period in the last 3,700 years." We also showed that "the growth surge occurred only within about 150 meters of upper tree line, regardless of tree line elevation."

This did not fit LaMarche's idea. The dramatically better mountain weather data available to us improved understanding of the processes controlling ring growth and led us to conclude that "increasing temperature at high elevations was likely a prominent factor" in the growth surge at these sites, not carbon dioxide fertilization. This opened the way to use the common variation of ring width in bristlecone at the highest elevations across the Great Basin, combined with changing upper tree-line elevation, to track summer temperature in this region over 4,500 years. The general pattern was of a long, slow decline of about 2 °F (1.1 °C) marked by irregular pulses of cooling likely associated with the climate effects of volcanoes, culminating in the unique twentieth- to twenty-first-century rapid warming. A detailed comparison with state-of-the art climate models confirmed the role of the greenhouse effect in causing this warming.

Cold snaps with global reach

As he explored the rings of bristlecone near the upper tree line, LaMarche, along with climatologist Katherine Hirschboeck, noticed characteristically damaged "frost rings" where early or late frosts had disrupted tracheid growth either late or early in the growing season. Especially in the most recent several centuries, they found a remarkable consistency in the occurrence of frost rings in high-elevation trees across a huge region. Frost rings were known to be produced by at least two cold nights (23°F / −5°C or worse) separated by a day also close to 32°F (0°C). To produce the characteristic damage, tracheids must still be forming, before their rigid wall structure is completed.

Looking for "cold snaps" across much of the western United States during the period of instrumental weather data led them to September 1884 and September 1965. In the tree rings of both these years in two widely separated mountain ranges—California's White Mountains and eastern Nevada's Snake Range—they found clear frost damage in the last part of the ring. Cool summers meant tracheid growth likely continued well into September, making it vulnerable to cold snaps. Even more intriguing was that those cool summers and cold snaps were linked to the climate effects of two major explosive volcanic eruptions in Indonesia, near the equator: Krakatoa in 1883 and Mount Agung in 1963–1964. To rule out coincidence, LaMarche and Hirschboeck tallied the dates of major explosive volcanic eruptions and those of widespread frost rings in western North America. Frost rings often appeared to occur within a couple of years after such an eruption, markedly more than by chance. This was especially the case in recent centuries for which the volcano dates were best known. LaMarche and Hirschboeck had made a good case for bristlecone frost rings as possible recorders of the weather effects of explosive volcanic eruptions of the kind that affect climate globally.

Two decades later, we went a step further. Really small or missing rings occurred at the lower elevations, for example MWK, in dry years. At the highest elevations in the same mountain ranges they could be produced either by drought or cold. We used five high-elevation millennium-length site chronologies in eastern California, across Nevada, and in northern Arizona—a huge region. From the lower-elevation site chronologies, we identified the drought years for each region and removed them from the list of very small or missing rings in their neighboring high-elevation site chronologies. This allowed us to identify 165 low-growth years over the past 5,000 years most likely caused by extreme cold events. Many of these rings were frost rings. Many also fell within a year or so of known explosive volcanic eruptions or of the acidity peaks they produced in annual layers in cores drilled through the Greenland Ice sheet.

These findings then contributed in 2015 to a revision of the global history of climatically effective explosive volcanic eruptions over the past 2,500 years. Combining ice-core, historical, and tree-ring data, the researchers reached the important conclusion that: "large eruptions in the tropics and high latitudes were primary drivers of interannual-to-decadal temperature variability in the Northern Hemisphere during the past 2,500 years." Think about that—a major contribution from these trees improved our knowledge and understanding of an important aspect of climate, far beyond the mountains where they grow, and was only possible with good chronology. Do note that, to reach this point from LaMarche and Hirschboeck's work took thirty years. Science takes persistence.

Sharpening the focus in time— from climate to weather events, from rings to cells

Dendrochronology has accelerated in the twenty-first century. It is now possible to measure the internal structure of thousands of tree rings, at the level of individual tracheids. Recent exciting results show that single-year, summer season temperatures can be got, also from near-tree-line bristlecone rings, by using a microscopic variant of medical computer tomography (CT) to measure the density of the last couple of cells formed in each ring—zooming in to ~20μm to look back thousands of years!

In the early 2010s, Alma Piermattei showed how the staining of ~20μm thin microscope cross-sections could be used to check how complete tracheid wall formation was in bristlecone. In the technique she used, a completely formed tracheid wall stains red, but if the process of filling up the matrix of the wall with lignin is not complete, it stains bright blue. The process, called lignification, is temperature dependent. So, "blue rings," where the lignification process is incomplete, are a record of low temperatures.

Sharpening the focus in space—
from sites to individual trees

As with blue rings and frost rings, the precise position of a tree on the Great Basin's rugged mountainsides exposes it to a remarkably different microclimate from a neighbor merely several feet away. Within a few feet, the steepness of slope and its compass direction, along with whether on a converging spot (land sloping toward) or not (land sloping away), can strongly affect the daily and seasonal march of temperature, drainage, and wind exposure. It was Andy Bunn's foxtail pine work in California's Sierra Nevada that alerted us to this possibility, and I eagerly encouraged him to bring his approach to the bristlecone. He and his students have

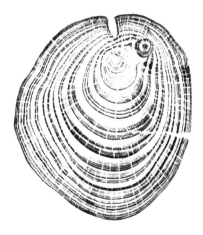

brought meter-scale topography and its relevance to tree growth, especially ring formation, into sharp focus. The occurrence or not of frost rings and blue rings in trees on these different facets of the mountainside hinted that quite different tree-ring records, for example of ring width, exist there too, within several feet of one another. That has turned out to be true. Not only will this help better interpretation of tree rings from long ago if we have precise locations but it will be of great value in thinking how warming climate is affecting the growth of the bristlecones in their rocky, irregular homes.

Coming full circle—
bristlecones and the Solar System

In the 1970s, while Wes Ferguson was pushing back the MWK chronology to almost 9,000 years before present, in Northern Ireland, researchers were also working back almost 8,000 years, using thousands of oak samples, from living trees in the present, through building timbers in recent millennia, to bog oaks before that. Just as Ferguson had to defend the MWK chronology, so the Belfast group had to overcome skepticism about crossdating working in such an "equable climate." I played a small part in explaining the mechanism that made crossdating possible in those oaks.

Why did this matter? To rely on any dating method, for example dendrochronology, we need to understand how and why it works. Similarly, radiocarbon dating works because the radioactive isotope of carbon, C14, is produced by cosmic rays bombarding the nitrogen that makes up most of Earth's atmosphere. That carbon, as carbon dioxide, is captured by green plants and turned into organic matter, living and then dead. From that point on, the proportion of carbon atoms in it produced by this bombardment should not change, except by radioactive decay. The rate of decay is known from physics and allows the calculation of the half-life of C14, which is about 5,730 years. If the concentration of C14 in the atmosphere is stable, the age of a piece of organic matter can be calculated from the proportion of C14 atoms in its carbon. Ferguson's accurately and precisely tree-ring dated pieces of bristlecone wood provided a way of checking this.

In fact, radiocarbon dates for material from most of the second millennium CE were found to be too old, and before that, increasingly too young until in the fifth millennium BCE when they are about 800 years too young. A few years later, this same pattern was seen in the Irish oaks, suggesting this is a global effect. It is believed to be the product of changes in Earth's magnetic field that shields, to a varying degree, the atmosphere from the cosmic ray bombardment that produces C14. Early in the development of the radiocarbon calibration, the smallest pieces of wood that were usable covered ten or twenty years. The coarseness of the C14 counting then available limited the detail of the calibration to this grand sweep across the past 9,000 years. Even so, this produced major upheavals in knowledge of the human and environmental past.

Wiggles with global effects

Over the decades, as the technology improved, the amount of wood needed decreased, right down to being able to measure C14 in a single year's ring, and the detail of the calibration curve improved dramatically. With the MWK bristlecone and then extending to Ireland and western European oaks and far beyond, many fine details of rises, falls, and plateaus in the comparison of radiocarbon and tree-ring dates have been shown to be global. They are related to variations—"wiggles"—in the Sun's behavior and so the bristlecone has now also contributed to solar physics. These wiggles over a few years in turn play an important role in the use of radiocarbon dating.

MALCOLM K. HUGHES

Now the drive is on to check radiocarbon year by year for several millennia at multiple locations around the planet. This is motivated by the 2012 discovery of an extreme radiocarbon spike in a single year, 774 CE, in Japanese cedar tree rings by Fusa Miyake and her colleagues. First thought to be the product of a cosmic ray burst from a supernova, it turned out to be produced by a major solar storm. This storm must have bombarded Earth with solar energetic particles that today would strongly disrupt much modern infrastructure, including communications, navigation, power grids, and space satellites. That same C14 spike is found in trees in several places around the globe, all caused by the same extreme solar storm. Of course, the bristlecone at MWK are an important resource for this search. At the time of writing, five clear "Miyake" events are known, with four more possibles under investigation, and probably more to be uncovered as the year-by-year radiocarbon analyses proceed.

As well as stimulating research in solar physics and its possibly serious implications for society, the discovery of these events is providing global marker years for the chronology of organic materials. Once again, seven decades on from Schulman and his "living ruins," the Great Basin bristlecone pine is helping us keep time on the events and processes that continue to shape our natural world and human society. Those scrubby, zombie trees, sitting crooked on their rocky slopes among fallen neighbors, are a gift that has kept on giving. We must take care of them to ensure they can continue to do so.

. . .

10
...
LAÑILAWAL

JONATHAN BARICHIVICH-HENRÍQUEZ

SPECIES
Alerce Cypress, *Fitzroya cupressoides*
....
LOCATION
Los Ríos and Los Lagos districts, Chile
....
ESTIMATED AGE
Up to 5,000 years old

There are about three trillion trees across the forests of the planet. Some of these trees are exceptional because they can live for thousands of years. Yet very few can reach 4,000 to 5,000 years, and even fewer of them have survived the destruction brought by the Anthropocene. Among this handful of surviving ancient trees is Lañilawal, an exceptional individual of alerce cypress (*Fitzroya cupressoides*) growing in the cool temperate rainforests of southern Chile. This one remarkable tree holds a deep connection to my family and has profoundly shaped my journey as an Earth scientist. As this tree's human ambassador, I am committed to its protection and to raising global awareness about the importance of understanding and safeguarding the last remnants of ancient trees and forests on the planet.

Alerce is one of a select few species whose enduring presence over millennia allows them to witness and record Earth's grandest cycles and rare cataclysmic events.

Since woody trees first appeared around 380 MYA, tree species have come and gone as continents shifted and the climate and the biosphere evolved. The southern alerce cypress, or *lawal*, diverged from its closest relatives approximately 30–40 MYA when South America was separating from a freezing Antarctica and drying Australia as part of the final breakup of the supercontinent Gondwana. Since its early days, the species has persisted in ever-wet, cool temperate environments that have shifted, expanded, and contracted dramatically over time in regions that today correspond to northern Patagonia in southern South America. Its evolutionary history has paralleled the uplift of the Andes, which significantly influenced its habitat and distribution. Because of its ancient Gondwanan heritage and wider distribution in the past—similar to several other conifers of the Southern Hemisphere—alerce is considered a paleoendemic or relict conifer now restricted to a much smaller, isolated geographic area.

At the rhythm of the ice ages

In more recent geological times, the glacial cycles of the Pleistocene had a profound impact on the habitat and biogeography of alerce within its already narrow range. As ice sheets advanced northward and descended from the Andes, populations also retreated northward, toward the lowlands of the Central Depression and the foothills of the Chilean Coastal Range, and possibly to some lowlands east of the Andes. During the

JONATHAN BARICHIVICH-HENRÍQUEZ

warmer interglacial periods, the species recolonized new areas as the ice receded toward the south. These cycles of range contraction and expansion shaped the current disjunct distribution of the species both in the temperate Andes and on the summits of the Coastal Range in south-central Chile, leaving a lasting mark on its genetic diversity.

Living at cosmic frequencies

Unlike any other living organism, ancient trees live closer to the rhythms of cosmic time. Only alerce and a select few other species such as bristlecone pines and giant sequoias attain lifespans measured in millennia, connecting their present to a deep and distant past. Their enduring presence allows them to witness and record Earth's grandest cycles and rare cataclysmic events. Before human activity began shaping the remote southern forests, alerce trees often lived to their full potential, reaching colossal sizes if they managed to escape catastrophic volcanic eruptions or occasional wildfires. Around 12,800 years ago, alerce trees likely witnessed a catastrophic cosmic event—the impact of a disintegrating meteorite, the Younger Dryas comet, that is believed to have caused widespread defaunation and climate change across the Americas, including the lands where alerce thrived.

Early human encounters

Humans may have reached the coastal regions of southern South America, where alerce occurs, as early as 14,500 to 18,500 years ago. These early settlers encountered forests dominated by ancient trees, untouched by human influence. Such first encounters initiated a complex and transformative relationship between people and the once-wild southern lands. Direct evidence of alerce use by these hunter-gatherers comes from the archaeological site of Monte Verde, near the coastal city of Puerto Montt. Artifacts from this site suggest that populations living in cold, postglacial environments utilized alerce wood for domestic purposes, hut construction, and tool-making at least 12,500 years ago.

LAÑILAWAL TIMELINE

• • •

Key milestones in the timeline of the alerce
and Lañilawal tree-ring record

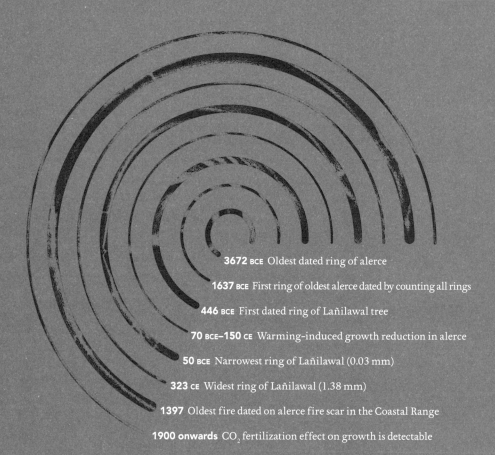

3672 BCE Oldest dated ring of alerce

1637 BCE First ring of oldest alerce dated by counting all rings

446 BCE First dated ring of Lañilawal tree

70 BCE–150 CE Warming–induced growth reduction in alerce

50 BCE Narrowest ring of Lañilawal (0.03 mm)

323 CE Widest ring of Lañilawal (1.38 mm)

1397 Oldest fire dated on alerce fire scar in the Coastal Range

1900 onwards CO_2 fertilization effect on growth is detectable

As human populations expanded and adapted over millennia, the cultural landscape of southern South America evolved along with the warming climate and shifting environments. Around 5,000 years ago, the ancestors of Mapuche people began to emerge as a genetically distinct population in what is now south-central Chile and parts of Argentina. Over time, they developed into the Mapuche culture we recognize today, characterized by a profound spiritual connection to the forests of the region, particularly with the araucaria and alerce forests.

For the Mapuche, forests are sacred spaces where the spiritual and physical worlds coexist. Animals, plants, rivers, wind, mountains, and humans all have a spirit, and these spirits are all interconnected. The Mapuche concept of *Itrofillmongen* embodies the interconnectedness and balance of all material and spiritual forms of life in coexistence. Itrofillmongen can be seen as the Mapuche equivalent of biodiversity, but it extends into the spiritual realm, encompassing the unseen forces that govern nature. Central to this balance are the *Ngen*, powerful guardian spirits that inhabit and protect specific elements of nature. Each Ngen is responsible for a particular domain, such as water (*Ngen Ko*) or alerce forests (*Ngen Lawal*). These spirits ensure that the Itrofillmongen is in balance.

The Mapuche people who coexisted with the alerce rainforests in their southern territories or *Butahuillimapu*, including the Chiloé Archipelago, were known as the Huilliche, or "people of the south." The Huilliche used the wood from alerce forests at lower altitudes near the coast for crafting utensils, tools, and weapons for defense. Their lands and way of life remained free of European influence until the mid-sixteenth century, when the Spanish invaders entered the Butahuillimapu and established the first cities. Spanish exploration of these southern territories began under campaigns of Pedro de Valdivia, who in 1552 founded the city of Valdivia near the northern range of the coastal alerce forests. Shortly after, in 1567, the Spaniards established Castro, the capital of Chiloé Island, marking one of the first permanent European settlements in proximity to the core region of alerce forests in the Andes.

Disastrous encounters with settlers and colonizers

The Spanish did not find gold in Chiloé, but instead discovered a rich mine of alerce wood. They quickly recognized its unique durability and extreme water-resistance, and began exploiting the ancient alerce forests to construct buildings and infrastructure in their newly founded cities. Alerce shingles became a defining architectural feature of the southern settlements, lending a distinct character to the towns and cities of the region. On Chiloé Island, these wooden shingles were essential to the construction of its iconic churches, many of which are now celebrated as UNESCO World Heritage sites. The wood of alerce itself became a form of currency within the emerging colonial economy of the province. The demand for alerce soon extended beyond the region, as a lucrative trade in alerce wood emerged, including exports to Peru. By the late eighteenth century, however, intense exploitation had severely depleted the more accessible forests in the lowlands closer to the cities and along the Andean shoreline. The logging areas started to progressively move farther and upward into the mountains, and by then they were owned by the king of Spain.

JONATHAN BARICHIVICH-HENRÍQUEZ

The method of free and selective exploitation of alerce forests during the colonial centuries changed dramatically following the establishment of the Republic of Chile in 1818. Private land ownership was formally established. This prompted a surge in often fraudulent land purchases from Mapuches, auctions, and large-scale concessions, particularly to foreign companies, fostering land speculation and facilitating indiscriminate exploitation of the forests. By the mid-nineteenth century, the Chilean government actively promoted large-scale colonization of the territories between Chiloé and Valdivia, with German settlers burning vast tracts of densely forested land to convert it into arable land for agriculture and pasturelands. The forests, once home to alerce trees of extraordinary size of up to 13 ft (4 m) in diameter, were rapidly exterminated, leaving behind only scattered stumps and some trophy photographs that are today silent witnesses to their former grandeur.

After four centuries of exploitation, only around 40 percent of the original pre-Hispanic alerce forest area remains. In the 1970s, alerce received national and international protection as an endangered species. Today, several national parks safeguard alerce forests; however, much of what remains today consists of secondary forests, regenerated after repeated fires and multiple cycles of exploitation. Despite these measures, many alerce populations in Chile remain at risk of extinction. Illegal logging continues to threaten the species, driven by the high market value of its exceptionally durable timber. Road construction, including projects cutting through the Alerce Costero National Park, and urban development further encroach on its habitat, fragmenting already vulnerable forest areas. Intentional wildfires, exacerbated by climate change, pose an additional and growing threat.

Naturalist identity

Known as *Lawal* or *Lahual* by the Huilliche people, the species was named alerce by the Spanish colonizers, since it reminded them of the European larch. Early accounts of British explorers passing through the west coast of South America in the late eighteenth and early nineteenth centuries noticed the species owing to the quality, beauty, and value of its wood.

In 1782, the Chilean Jesuit botanist Juan Ignacio Molina provided the first formal scientific description of the species, naming it *Pinus cupressoides*. Despite this, the species remained poorly known in British botanical circles well into the nineteenth century. Captains Robert Fitzroy and Phillip King, in their narratives of the HMS *Beagle* and *Adventure* voyage published in 1839, described the use of alerce for shingles and boards, noting trees measuring up to 30 ft (2.9 m) in girth in the mainland Chiloé region. Fitzroy speculated that even larger specimens reaching diameters up to 13 ft (3.9 m) might exist in the Cordillera out of reach of local woodsmen.

Charles Darwin, who accompanied Fitzroy on the voyage of the *Beagle*, visited Chiloé in 1834 and explored the Chilean rainforests. However, Darwin himself admitted to having little botanical knowledge. In a letter to the botanist Joseph Dalton Hooker in 1843, he confessed, "From my entire ignorance of Botany, I am sorry to say that I cannot answer any questions which you ask me. I saw the Alerce on mountains of Chiloe (on the mainland it grows to an enormous size, and I always believed Alerce and *Araucaria imbricata* to be identical), but I am ashamed to say I absolutely forget all about its appearance."

After several revisions, the final scientific name, *Fitzroya cupressoides*, was assigned, in honor of Captain Robert Fitzroy, possibly more as a gesture of admiration than as a recognition of any significant contribution to documenting the tree. Alerce did not share the fortune of its northern coniferous neighbor, *Araucaria araucana*, whose scientific name honors the Arauco region and its Indigenous people. Perhaps the scientific name of alerce should be revised in the future to free it from botanical colonialism and better reflect its true cultural heritage. A name such as *Lawalia chiloensis* or *Lawalia mawidensis* would better embrace the ancient identity bestowed by the Huilliche people.

An unconventional southern ecology

The long-lived rainforest conifers of the Southern Hemisphere have historically defied conventional ecological understanding due to their exceptional lifespans and slow growth, with alerce being an extreme case. Their life cycles unfold in rhythms alien to the temperate paradigms of Europe that long ruled ecological thinking. By the mid-twentieth century, four centuries of relentless exploitation and repeated burning had changed the geography and ecology of the alerce forests forever. Foresters and biogeographers of the time observed a startling absence of seedlings and young regeneration in many forests, especially in areas heavily logged and burned or shaded by dense canopies. This apparent inability to regenerate naturally, combined with its slow growth, gave rise to a prevailing belief: Alerce was a living fossil, a relict of a colder and wetter past climate. Under this view, it was destined to natural extinction or to be displaced to infertile sites by faster-growing broadleaved species of the Valdivian rainforest. This paradigm embodied the earlier ecological misunderstanding of alerce, casting it as a species out of pace with the modern natural world.

Alerce, however, is a shade-intolerant species that relies on open spaces for its regeneration. Such open spaces are mostly created by generally rare natural and anthropogenic disturbances, such as volcanic eruptions, landslides, windthrow, fire, and logging. In old-growth forests, alerce's extreme longevity is a strength, because it allows the species to persist for centuries while waiting for rare disturbance opportunities to regenerate. In the early 1990s, Chilean forest ecologist Antonio Lara

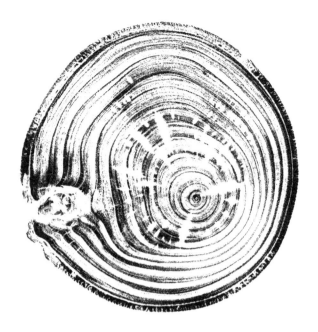

tested this disturbance hypothesis and found that in the Andes, natural disturbances can be highly sporadic, leading to gaps of up to five to six centuries between recruitment events.

This clarified the observed lack of regeneration in virgin closed-canopy forests and the profuse regeneration following natural disturbances and low-intensity fires. While natural regeneration in heavily logged or burned areas is exceedingly slow and often dominated by faster-growing species in the short term, the remarkable lifespan of alerce enables it to recover over centuries if seed sources or isolated surviving trees remain. Ultimately, the paradigm of alerce as a climatic relict was overturned by recognizing that slow growth and longevity are the very cause of its resilience rather than its demise.

With its slow growth and millenary longevity, alerce challenges another mid-twentieth-century prevailing paradigm, this time of North American origin. According to the "longevity under adversity" paradigm, maximum longevity is thought to occur only in harsh, dry, and cold environments like those found in the mountains of the Southwest in the United States. In 1950, Edmund Schulman, who also found

JONATHAN BARICHIVICH-HENRÍQUEZ

the long-lived bristlecone pines, collected alerce samples throughout Chile and Argentina, including a 1,800-year-old tree from Argentina and a 1,600-year-old tree from Chile. Based on the "longevity under adversity" paradigm, Schulman expected to find the oldest trees and the narrowest rings in the arid, cool regions of his collection. But to his surprise, he found record-low growth rates in the healthy, 1,600-year-old Chilean alerce, which he sampled in a temperate, extremely wet environment with more than 118 in (3,000 mm) rainfall per year. Why and how this occurs remains an enduring ecological mystery until today, shrouded in scientific questions we have yet to answer or even pose.

Opening an ancient time capsule

The very slow growth and extreme longevity of alerce trees have long been acknowledged. To the naked eye, it is evident from the tightly packed growth rings in stumps and boards, and maximum diameters of 10 to 16½ ft (3 or even 4 m). A popular myth was that alerce trees grow about 1 mm in diameter per year. It was a simple yet often misleading way for early observers to estimate the age of large trees. While this rule of thumb may have frequently led to overestimations due to varying growth rates and the faster growth during early life, it fits well with a reliable early ring count from a felled alerce tree in Argentina in 1956. The tree was 10 ft (3 m) in diameter and had 3,000 annual rings.

Yet, the first crossdated and well-replicated alerce tree-ring chronology was not developed until 1985, after which research into alerce tree rings advanced significantly in the 1990s and 2000s. These advances were led by dendrochronologists Ricardo Villalba in Argentina and Antonio Lara in Chile.

I first met Antonio in the summer of 1993, during his visit to the nowadays Alerce Costero National Park, in the summit of the Coastal Range near Valdivia, to study the fire history of these enigmatic alerce forests. As an intrepid eleven-year-old kid who grew up in the park with my family of rangers, I guided Antonio and his colleagues to alerce stands suitable for study. Antonio showed me how the dates of ancient fire scars were recorded in the eroded growth rings of unassuming alerce stumps. From that moment on, I could never look at a stump or an alerce tree the same way again—each one became a time capsule, holding blurry messages from the past. This unexpected encounter became a defining event in shaping my scientific journey, a path that probably began when I took my first step among ancient alerce trees, holding my mother's hand under the watchful presence of my grandparents.

Six years later, as I stood at the crossroads of choosing a possible university career, Antonio encouraged me to study forestry at the Universidad Austral de Chile in nearby Valdivia, where he had recently founded the country's first tree-ring laboratory. However, the tuition fees were far beyond my means. Moved by my particular family story with the alerce forests, the director of the forestry school helped me to obtain a highly competitive full scholarship. It felt as if the very ancient trees that had shaped my childhood were now secretly guiding my destiny.

During my busy student life, I began dating alerce tree-ring samples in the lab late into the night. The quiet of those hours allowed me to immerse myself into learning the intricate patterns of the language of ancient trees. From time to time, Antonio and I would cross each other in the corridor, the only two people in the building at that late hour, or he would stop by to see who was still working in the lab. Though brief, these moments always sparked inspiring conversations, where we recognized in each other a shared scientific passion for looking into the secret messages hidden inside the time capsules of ancient alerces.

Antonio went on to co-supervise my undergraduate thesis, where I used tree rings to understand the dieback of alerce trees that I knew in the very park where I had grown up. This work was an early investigation into a phenomenon that has since become increasingly prevalent around the world due to climate change. For me, it was a first glimpse into the unconventional ecology and the fragility of these ancient trees.

The collaborative efforts of Antonio and Ricardo led to the creation of a network of over twenty ring-width chronologies spanning the entire range of the species and several millennia, some of which

I helped to collect, and to the study of alerce tree rings as a multi-millennial record of summer temperatures. Before retiring in 2023, Antonio used Chilean alerces to reconstruct 5,680 years of summer temperatures for the southern Andes (3672 BCE to 2009). Today, this stands as the longest tree-ring record for the Southern Hemisphere.

From ancient forests to global models

After a PhD and post-doc in Europe, where I complemented my earlier tree-ring experience with earth system model expertise to study global forest ecology, I returned to Chile to complete the retribution time required by the scholarship that funded my PhD. What first felt like an interruption to my scientific career ultimately proved to be a meaningful call back home from the ancient trees that had guided my path from the start.

Two years before my return, Antonio had launched a pioneering project to measure carbon and water fluxes in an alerce forest using an eddy covariance flux tower in the Alerce Costero National Park, where we had first met twenty years earlier. After an unfortunate lightning strike left the costly tower inoperable, my return became an opportunity to reconnect with Antonio and restart the project. We joined forces with my longtime friend and colleague, Rocío Urrutia, whose research had revealed that alerce forests are among the most massive and long-lived carbon stores in the world, with the highest wood residence times reported for any species. On average, a molecule of carbon captured by photosynthesis and stored in alerce wood takes 1,400 years to return to the atmosphere through decomposition. As a result, alerce forests are unique long-term carbon sinks due to their exceptionally low mortality rates and the ability to slowly accumulate highly durable biomass for millennia.

I returned to my family land near the park, living with my mother just a short walk from the flux tower to manage its complex operations. Immersed once again in the alerce forests of my childhood and removed from the scientific hustle of Europe, I found a unique space of inspiration to mature my scientific vision. Measuring the pulse of the ancient alerces became a family affair. My mother monitored cambial activity and shoot elongation of alerces on a weekly basis, continuing a long family tradition of supporting scientific research in the park, but this time the scientist was at home.

To me, the time spent in the forest was not just about data collection, but about reconnecting with the source of inspiration of my career. The quiet, immersive presence in the ancient alerce forests allowed me to reflect on how the detailed, traditional methods of studying tree rings could complement the real-time, ecosystem-level data from the flux tower. The synergy between the deep past and present gave me a fresh perspective on how we might better predict the future. Through the mysterious language of inspiration, the alerces revealed to me that this could be achieved through a scientific model that weaves together the depth of historical tree-ring data with the biophysical pulse recorded by continuous forest monitoring. This idea became the foundation of a successful career-starting grant proposal, which took me back to Europe, this time as a senior scientist.

The encounter with Lañilawal

In the early 1970s, my grandfather, Aníbal Henríquez, came across a massive, ancient alerce tree while working as the first ranger of the Alerce Costero National Park. This unexpected encounter marked the beginning of a profound connection between my family and this tree, the largest and oldest living individual in the rainforest. Initially, he chose to keep the finding to himself, reluctant to reveal the location of the tree. A few years later, he shared the news with the rest of the family and the park manager. He eventually cleared a small trail to the tree, which became not only the main attraction of the park but also one of the most iconic trees of the Chilean rainforest and the world.

The tree is commonly known as "Alerce Milenario," but its true name is *Lañilawal*, a Mapuche word of uncertain meaning. This mysterious name connects the tree to the *Lañilawal* river at the bottom of the valley and the broader material and immaterial world it inhabits.

JONATHAN BARICHIVICH-HENRÍQUEZ

While the steam from a
hot cup of tea rose in the kitchen, outside
the wind and rain had not stopped for days. But
once the sun came out, it all ceased. The damp sound of the
forest whispered to Aníbal, you have to go. He had to find it. It
had been waiting for thousands of years. Aníbal wasn't quite sure
how he found the right path, but an ancestral compass led him. There it
was, as it had always been. Silent, unnoticed in the landscape. It had
endured tempestuous winds and snowstorms. It had sheltered birds, and
plants and even trees had grown on it. When Aníbal saw it, he couldn't believe
it—it was the ancient Lañilawal. Aníbal wondered, why him. Lañilawal
answered, you will be the guide and protector, and you will teach the
importance and fragility of all that inhabit here. Aníbal fulfilled his promise.
He took his grandson's hand and showed him every path and plant, with
simple gestures and words. What seemed impossible had finally
happened. Aníbal departed, and his grandson followed in his
footsteps, going further. Now he climbs a tower, measures the
wind, and listens to the whispering of the trees. He gathers
scientific data and speaks in foreign languages.
He says that the paths of life are shaped
by loving the past.

"THE ENCOUNTER"
JUAN DIEGO MALDONADO, SANTIAGO, CHILE, 2021
....

Beyond being an ancient living being, we consider it to be a *Ngen Lawal*, a powerful guardian spirit of the alerce forests of all times. Over three generations of park rangers, my family—beginning with Aníbal and my grandmother Rufina, and then joined by my mother Nancy, uncle Jorge, and cousins Marcelo and Deborah—has protected Lañilawal during half a century. Today, the legacy of safeguarding it has been passed to me, not only as a descendant but as an Earth scientist, committed to ensuring its wellbeing, freedom, and survival for the centuries to come.

When my grandfather first met Lañilawal, it was a marvel of untouched nature, at nearly 14 ft (4.2 m) in diameter and 92 ft (28 m) in height. The massive tree was draped in a living cloak of mosses and epiphytic plants, so dense that even a large tree had taken root on top of one of its branches. This intricate ecosystem gave Lañilawal a mythical appearance, like a cascading waterfall of lush green and brown hues that camouflaged it within the rainforest. However, as its fame grew, so did the pressure from uncontrolled tourism. Even after a wooden viewing platform was constructed to protect it, visitors easily trespassed, trampling its last living roots, compacting the soil, and suffocating the tree. They peeled away its bark and stripped all the mosses and plants that once adorned it. Decades of neglect and overtourism took their toll, and by the time I returned to Chile from Europe in 2016, Lañilawal stood in a diminished state, like a caged lion, stripped of its former wild majesty.

Examining Lañilawal

When I saw Lañilawal in such a poor condition, it was clear that overtourism had left it vulnerable. Compounding the issue, I noticed signs of climate change stress affecting the surrounding forest, including dying trees, desiccated mosses, and the mortality of high epiphytes, once permanently soaked in moisture. It was a worrying sight of how rapidly the vital space around the tree and the surrounding ecosystem were deteriorating. In the winter of 2019, with these concerns in mind, together with Antonio Lara, we decided to carefully extract a tree-ring sample from the ancient Lañilawal to evaluate its health condition.

Accompanied by the park managers and rangers, we approached the task with utmost care, ensuring the process was respectful and minimized any potential harm.

The analysis of the tree-ring sample confirmed the external signs of vulnerability of Lañilawal to increased dryness and warming. It showed that its radial growth increased steadily with rising concentrations of carbon dioxide in the atmosphere during the twentieth century, but this growth stimulation halted in 1990 as atmospheric dryness and maximum temperatures rose in the region. With climate change projections indicating further increases in dryness and warming, the vulnerability of Lañilawal and the surrounding ecosystem is likely to intensify, threatening its survival.

Looking toward the future, with the help of my brother Diego, we started monitoring the immediate habitat of Lañilawal in 2021, measuring soil and air temperature and moisture every half hour. These measurements are integrated with the broader monitoring of the alerce ecosystem at our flux tower, located just 1½ miles (2.5 km) away. The data revealed a concerning issue: The large wooden viewing platform, installed just 6½ ft (2 m) from the trunk of Lañilawal and directly above its last living roots, acts as a barrier to precipitation, reducing soil moisture to nearly half the levels found in the surrounding forest. Rather than protecting the tree, the platform inadvertently intensifies drought stress. We have urged park authorities to promptly relocate the platform at least 50–65 ft (15–20 m) away from the tree to minimize disturbance and better preserve its growing environment as droughts intensify with climate change.

Age of Lañilawal

For many years, a popular belief circulated that this ancient alerce was around 3,500 years old, a claim displayed on a wooden information sign beside the tree. However, the origin of this estimate is unknown. It is likely based on the popular belief that alerces grow in diameter at a rate of approximately 1 mm per year. With a trunk diameter of nearly 14 ft (4.2 m),

an average growth rate of 1.2 mm per year over bark would have to be assumed to arrive to the age estimate of 3,500 years. However, estimating the age of Lañilawal using traditional tree-ring counting is challenging because of its immense size. Doing it was not the purpose of our sampling. However, after finding 2,465 tightly spaced rings within the 3-ft (90-cm) core we sampled, we decided to explore the possibility of estimating the total age. This sample, though significant, represents only 43 percent of the tree trunk radius. The scientific challenge then became how to estimate the rings in the remaining 57 percent of the trunk radius to provide a more complete age estimate with its uncertainty.

To do this, we developed a statistical growth model to estimate thousands of possible growth trajectories from seedling to adult, consistent with measurements for the species and the tree itself. We found an 80 percent chance that Lañilawal is more than 5,000 years old, suggesting it may be one of the oldest living, non-clonal trees in the world. This preliminary finding was highlighted in the scientific journal *Science* in 2022, attracting worldwide attention but also skepticism from some North American dendrochronology colleagues who believe that tree age must be determined solely by counting all the rings. We are in the process of publishing our methodology, but for now, our focus has been directed toward other aspects of the tree that required more urgent action.

JONATHAN BARICHIVICH-HENRÍQUEZ

While these findings offer a plausible age estimate, the exact number holds little significance when compared to the true importance of Lañilawal. The focus should not be on the age of this ancient being, which is merely a number for humans to measure. Instead, we should turn our attention to the deeper meaning of its existence at a cosmic time—its enduring role as a guardian spirit, its connection to the land and the cosmos, and its continued resilience in the face of ever shifting challenges. The true value of Lañilawal lies not in its age, but in the messages it carries for all of us who are willing to listen.

Life continues

In its spiritual role as Ngen, Lañilawal chose to reveal itself to humanity in a timely act to protect itself from the destructive power of civilization. Fate chose my grandfather and family to be its human kin, a bond that has deepened over generations and inspired my scientific career. The human family of Lañilawal grows—it now extends to Antonio, Rocío, committed park rangers, and all those who stand in solidarity for its protection and freedom. But there will come a moment when Lañilawal must be left in peace, a time when it should not be visited, allowing it to heal and continue carrying its silent messages for the humans and spirits of the future. It must not be reduced to a mere tourism attraction, nor should its genome be desecrated, stealing its most intimate secrets before their time. There are frontiers that scientific colonialism must not cross to respect the cosmovision of its ancestral people. Instead of being commodified and exploited, it should be revered, protected, and understood. The story of Lañilawal and that of its species reminds us that despite adversity, life continues and the future is always uncertain.

· · ·

Gallery

Giant sequoia
Sequoiadendron giganteum

(PAGE 16)

Bosnian pine
Pinus heldreichii

(PAGE 38)

Clanwilliam cedar
Widdringtonia cedarbergensis

(PAGE 56)

Kauri
Agathis australis (D.Don) Lindl.

(PAGE 76)

Bald cypress
Taxodium spp. (L.) Rich.

(PAGE 94)

Pedunculate oak
Quercus robur L.

(PAGE 112)

IN THE CIRCLE OF ANCIENT TREES

Cedro
Cedrela odorata

(PAGE 132)

Qilian juniper
("祁连圆柏" in Chinese;
pronounced "*qí lián yuán bǎi*"),
Juniperus przewalskii Kom.

(PAGE 150)

Great Basin
Bristlecone Pine
Pinus longaeva

(PAGE 170)

Lañilawal
Fitzroya cupressoides

(PAGE 190)

IN THE CIRCLE OF ANCIENT TREES

FURTHER READING

Chapter One: Giant Sequoia

Harvey, H.T., H.S. Shellhammer, and R.E. Stecker. *Giant Sequoia Ecology: Fire and Reproduction*, National Park Service Scientific Monograph Series 12 (NPS, 1980)

Save the Redwoods League. "Giant sequoias and fire: The past, present and future of an ancient forest. An interactive story map" (2022), www.savetheredwoods.org/interactive/giant-sequoia-and-fire

Swetnam, T.W. "Fire history and climate change in giant sequoia groves." *Science* 262: 885–889 (1993)

Swetnam, T.W., C.H. Baisan, A.C. Caprio, et al. "Multi-millennia fire history of the Giant Forest, Sequoia National Park, USA." *Fire Ecology* 5(3):117–147 (2009)

Tweed, W.C. and L.M. Dilsaver. *Challenge of The Big Trees: The History of Sequoia and Kings Canyon National Parks* (George F. Thompson Publishing; revised edition, 2017)

Chapter Two: Bosnian Pine

Cook, E.R., R. Seager, Y. Kushnir, et al. "Old World megadroughts and pluvials during the Common Era." *Science Advances* 1: e1500561 (2015)

Panayotov, M. and N. Tsvetanov. "Dating of avalanches in Pirin mountains in Bulgaria by tree-ring analysis of *Pinus peuce* and *Pinus heldriechii* trees." *Dendrochronologia* 85 (2024)

Trouet V., M. Panayotov, A. Ivanova, et al. "A pan-European summer teleconnection mode recorded by a new temperature reconstruction from the northeastern Mediterranean (AD 1768–2008)." *The Holocene* 22: 887–898 (2012)

Vasileva, P. and M. Panayotov. "Dating fire events in *Pinus heldreichii* forests by analysis of tree-ring cores." *Dendrochronologia* 38: 98–102 (2016)

Xu, G., E. Broadman, I. Dorado-Liñán, et al. "Jet stream controls on European climate and agriculture since 1300 CE." *Nature* 634: 600–608 (2024)

Chapter Three: Clanwilliam Cedar

Dunwiddie, P.W. and V.C. LaMarche Jr. "A climatically responsive tree-ring record from *Widdringtonia cedarbergensis*, Cape Province, South Africa." *Nature* 286(5775): 796–797 (1980)

February, E.C. and W.D. Stock. "The relationship between ring-width measures and precipitation for *Widdringtonia cedarbergensis*." *South African Journal of Botany* 64(3): 213–216 (1998)

February, E.C. and W.D. Stock. "Declining trend in the 13C/12C ratio of atmospheric carbon dioxide from tree rings of South African *Widdringtonia cedarbergensis*." *Quaternary Research* 52(2): 229–236 (1999)

February, E.C., A.G. West, and R.J. Newton. "The relationship between rainfall, water source and growth for an endangered tree." *Austral Ecology* 32(4): 397–402 (2007)

Hall, M. "Dendroclimatology, rainfall and human adaptation in the later Iron Age of Natal and Zululand." *Annals of the Natal Museum* 22(3): 693–703 (1976)

Hubbard, C.S. "Observations on the distribution and rate of growth of Clanwilliam cedar *Widdringtonia Juniperoides* Endl." *South African Journal of Science* 33(3): 572–586 (1937)

Taylor, H.C. *Cederberg Vegetation and Flora* (National Botanical Institute, 1996)

White, J.D.M., S.L. Jack, M.T. Hoffman, et al. "Collapse of an iconic conifer: Long-term changes in the demography of *Widdringtonia cedarbergensis* using repeat photography." *BMC Ecology* 16: 1–11 (2016)

Chapter Four: Kauri

Boswijk, G. and M.J. Jones. "Tree-ring dating of colonial-era buildings in New Zealand." *Journal of Pacific Archaeology* 3(1): 59–72 (2012)

Boswijk, G., A.M. Fowler, J.G. Palmer, et al. "The late Holocene kauri chronology: Assessing the potential of a 4500-year record for palaeoclimate reconstruction." *Quaternary Science Reviews* 90C: 128–142 (2014)

Evans, K. "Buried treasure." *New Zealand Geographic* 142 (Nov–Dec 2016), www.nzgeo.com/stories/swamp-kauri

FURTHER READING

Fowler, A.M., G. Boswijk, A. Lorrey, et al. "Multi-centennial tree-ring record of enso-related activity in New Zealand." *Nature Climate Change* 2: 172–176 (2012)

Hogg, A., W. Gumbley, G. Boswijk, et al. "The first accurate and precise calendar dating of New Zealand Māori Pā, using Otāhau Pā as a case study." *Journal of Archaeological Science: Reports* 12, 124–133 (2017)

Orwin, J. *Kauri: Witness to a Nation's History* (New Holland Publishers; revised edition 2019)

Voosen, P. "Kauri trees mark magnetic flip 42,000 years ago." *Science* 371: 766–766 (2021)

Chapter Five: Bald Cypress

Delong, K.L., S. Gonzalez, J.B. Obelcz, et al. "Late Pleistocene baldcypress (*Taxodium distichum*) forest deposit on the continental shelf of the northern Gulf of Mexico." *Boreas* 50(3): 871–892 (2021)

Mattoon, W.R. *The Southern Cypress.* Bulletin of the U.S. Department of Agriculture, 272 (USDA, 1915)

Stahle, D.W., D.J. Burnette, J. Villanueva, et al. "Tree-ring analysis of ancient baldcypress trees and subfossil wood." *Quaternary Science Reviews* 34: 1–15 (2012)

Stahle, D.W., M.K. Cleaveland, D.B. Blanton, et al. "The lost colony and Jamestown droughts." *Science* 280: 564–567 (1998)

Stahle, D.W., J.R. Edmondson, I.M. Howard, et al. "Longevity, climate sensitivity, and conservation status of wetland trees at Black River, North Carolina." *Environmental Research Communications* 1: 041002 (2019)

Therrell, M.D., D.W. Stahle, and R. Acuña Soto. "Aztec drought and the 'Curse of One Rabbit.'" *Bulletin of the American Meteorological Society* 85: 1263–1272 (2004)

Villanueva Díaz, J., D.W. Stahle, B.H. Luckman, et al. "Potential for dendrochronology of *Taxodium mucronatum* Ten. and its conservation in Mexico." *Ciencia Forestal en Mexico* 32(101): 9–37 (2007)

Chapter Six: Pedunculate Oak

Daly, A., M. Domínguez-Delmás, and W. van Duivenvoorde. "Batavia shipwreck timbers reveal a key to Dutch success in 17th-century world trade." *PLoS One* 16: e0259391 (2021)

Daly, A. and I. Tyers. "The sources of Baltic oak." *Journal of Archaeological Science* 139: 105550 (2022)

Domínguez-Delmás, M., M. Driessen, I. García-González, et al. "Long-distance oak supply in mid-2nd century ad revealed: The case of a Roman harbour (Voorburg-Arentsburg) in the Netherlands." *Journal of Archaeological Science* 41: 642–654 (2014)

Domínguez-Delmás, M., S. Rich, and N. Nayling. "Dendroarchaeology of shipwrecks in the Iberian Peninsula: 10 years of research and advances." In: A. Crespo Solana, L.F. Monteiro Vieira de Castro, and N. Nayling (eds.) *Heritage and the Sea: Volume 2: Maritime History and Archaeology of the Global Iberian World (15th to 18th Centuries)* (Springer Nature, 2022) pp.1–57

Eaton, E., G. Caudullo, S. Oliveira, et al. "*Quercus robur* and *Quercus petraea* in Europe: Distribution, habitat, usage and threats." In: J. San-Miguel-Ayanz, D. de Rigo, G. Caudullo, T. Houston Durrant, and A. Mauri (eds.) *European Atlas of Forest Tree Species* (Publications Office of the European Union, 2016), pp.160–163

Haneca, K., K. Čufar, and H. Beeckman. "Oaks, tree rings and wooden cultural heritage: A review of the main characteristics and applications of oak dendrochronology in Europe." *Journal of Archaeological Science* 36: 1–11 (2009)

Jansma, E., *RemembeRINGs: The Development and Application of Local and Regional Tree-Ring Chronologies of Oak for the Purposes of Archaeological and Historical Research in the Netherlands* (Universiteit van Amsterdam/NAR 19, 1995)

Krąpiec, M. and P. Krąpiec. "Dendrochronological analysis of the Copper Ship's structural timbers and timber cargo." In: O. Waldemar (ed.) *The Copper Ship: A Medieval*

Shipwreck and Its Cargo
(The National Maritime Museum
in Gdańsk, 2014), pp.143–160

López-Bultó, O. and R. Piqué
Huerta. "Wood procurement
at the Early Neolithic site of La
Draga (Banyoles, Barcelona)."
Journal of Wetland Archaeology
18(1): 56–76 (2018)

Chapter Seven: Cedro

Granato-Souza, D., D.W. Stahle,
A.C. Barbosa, et al. "Tree rings
and rainfall in the equatorial
Amazon." *Climate Dynamics*
52: 1857–1869 (2019)

Granato-Souza, D.,
D.W. Stahle, M.C. Torbenson,
et al. "Multidecadal changes
in wet season precipitation
totals over the Eastern
Amazon." *Geophysical
Research Letters* 47(8):
e2020GL087478 (2020)

Muellner, A.N., T.D. Pennington,
A.V. Koecke, et al.
"Biogeography of *Cedrela*
(Meliaceae, Sapindales)
in Central and South
America." *American Journal of
Botany* 97(3): 511–518 (2010)

Chapter Eight: Qilian Juniper

Gou, X., Y. Deng, L. Gao, L.,
et al. "Millennium tree-ring
reconstruction of drought
variability in the eastern Qilian
Mountains, northwest China."
Climate Dynamics 45: 1761–
1770 (2015)

Liu, Y., Z. An, H.W. Linderholm,
et al. "Annual temperatures
during the last 2485 years in
the mid-eastern Tibetan Plateau
inferred from tree rings."
Science in China Series D: Earth

Sciences 52(3): 348–359 (2009)
Shao, X., S. Wang, H. Zhu, et al.
"A 3585-year ring-width dating
chronology of Qilian juniper
from the northeastern Qinghai-
Tibetan Plateau." *IAWA Journal*
30: 379–394 (2009)

Yang, B., C. Qin, A. Bräuning,
et al. "Long-term decrease in
Asian monsoon rainfall and
abrupt climate change events
over the past 6,700 years."
*Proceedings of the National
Academy of Sciences of the
United States of America*,
118(30): e2102007118 (2021)

Yang, B., C. Qin, J. Wang, et al.
"A 3,500-year tree-ring record
of annual precipitation on the
northeastern Tibetan Plateau."
*Proceedings of the National
Academy of Sciences* 111(8):
2903–2908 (2014)

Zhang, Q., G. Cheng, T. Yao,
et al. "A 2,326-year tree-ring
record of climate variability
on the northeastern Qinghai-
Tibetan Plateau." *Geophysical
Research Letters* 30: 1739
(2003)

**Chapter Nine:
Great Basin Bristlecone Pine**

Chiu, C. "Living ruins" (2024),
www.calvinchiuphotography.
com/blog (2024)

Cohen, M.P. *A Garden of
Bristlecones: Tales of Change
in the Great Basin* (University
of Nevada Press, 1998)

Lanner, R.M. *The Bristlecone
Book: A Natural History of the
World's Oldest Trees* (Mountain
Press Publishing, 2007)

McGraw, D.J. *Edmund
Schulman and the "Living*

*Ruins": Bristlecone Pines,
Tree Rings and Radiocarbon
Dating* (Community Printing
and Publishing, Bishop, 2007)

Schulman, E. "Bristlecone pine,
oldest known living thing."
National Geographic 113 (1958

Chapter Ten: Lañilawal

Gardner, M., T. Philip, A. Lara,
et al. "*Fitzroya cupressoides*."
Curtis's Botanical Magazine
16(3): 229–240 (1999)

Lara, A. and R. Villalba.
"A 3,620-year temperature
record from *Fitzroya
cupressoides* tree rings in
southern South America."
Science 260: 1104–1106 (1993)

Lara, A., S. Fraver, J.C. Aravena,
et al. "Fire and dynamics of
Fitzroya cupressoides (alerce)
forests of Chile's Cordillera
Pelada." *Ecoscience* 6(1): 100–
109 (1999)

Lara, A., R. Villalba, R. Urrutia-
Jalabert, et al. "A 5,680-year
tree-ring temperature record
for southern South America."
Quaternary Science Reviews
228: 106087 (2020)

Perez-Quezada, J.F., J.
Barichivich, R. Urrutia-Jalabert,
et al. "Warming and drought
weaken the carbon sink capacity
of an endangered Paleo-
endemic Temperate Rainforest
in South America." *JGR
Biogeosciences* 128(4):
e2022JG007258(2023)

Schulman, E. *Dendroclimatic
Changes in Semiarid America*
(University of Arizona Press,
1956)

INDEX

IN THE CIRCLE OF ANCIENT TREES

F

false rings 64, 139
Ferguson, Charles Wesley 179, 180, 187
fire scars 27–34, 36, 52–3, 202
Fitzroy, Robert 198–9
floating chronologies 109
Florissant Fossil Beds National Monument 18
Fowler, Anthony 78, 85, 86, 87, 88
foxtail pine 173, 178, 186
frost rings 184–5, 187
fynbos 59, 61, 63

G

General Grant 103
General Sherman 31
ghost forests 108
Giant Forest 32–3
giant sequoia 16–37, 52, 96, 103, 193
Gila Wilderness 27–8
Great Basin bristlecone pine 13, 170–89
Great Houses of Chaco Canyon 25–6
gymnosperms 7, 61, 62

H

haiduks 46
Hall, M. 62
Heldreich, Theodor von 46
Henriquez, Aníbal 204, 206, 209
Hirschboeck, Katherine 184, 185
Hooker, Joseph Dalton 198
Hubbard, C.S. 62
Hughes, Malcolm 32
Huilliche people 195, 198, 199
Huntington, Ellsworth 24–5, 29, 30
Hurricane Ivan 107

I

incense cedar 14
insect attacks 176, 178
International Council of Monuments and Sites 131
International Tree-Ring Database 142
Io Matua Kore 91
Itrofillmongen 195
IUCN Red List of Plants 74
ivory-billed woodpecker 106–7

J

Jamestown 100, 101–2
Japanese cedar 189
Jari River 138–9
jet stream 50–1
juniper, Qilian 150–69

K

kahikatea 82
Kang, Xingcheng 163
kauri 76–93
Khoekhoe 70–1
King, Phillip 198
Krakadouw 65, 66
Kruger, Fred 63, 72
Krusic, Paul 49
Kulakowski, Dominik 52

L

La Draga 119–20
La Niña 85–6
Laboratory of Tree-Ring Research 10, 12, 27
LaMarche, Val 63, 64, 65, 66, 68, 72, 181–4, 185
Lañilawal 190–209

Lanner, Ronald 175
Lara, Antonio 199–200, 201–3, 206, 209
larch, European 7, 14
Laschamps excursion 90
latewood 9, 117, 118, 126
Libby, Willard Frank 26–7
lignification 186
the *Limes* 121
Linnaeus, Carl 125
logging: alerce forests 196, 197, 199
    bald cypress 94, 109–10
    cedro 138, 146–7, 148, 149
    giant sequoias 24, 35
    kauri 83, 91
    pedunculate oak 120–1
Los Peroles 103–4, 106
Lost Colony 100, 101–2
Loudon, J.C. 84

M

McGraw, Donald J. 179
Majiayao culture 165
Maldonado, Juan Diego 205
Manaia Sanctuary 78, 81, 83, 86, 91, 93
Māori 83, 89, 90–1
Mapuche culture 195, 197
Marine Oxygen Isotope Stage 3 90
Martinez, Maximino 104
Maximino 104
Medieval Drought Period 33, 34
Methuselah Walk (MWK) 179, 180, 184, 187, 188, 189
Mirchev, Stefan 41
missing rings 161
Miwok culture 21, 37
Miyake, Fusa 189
Molina, Juan Ignacio 198
Monte Verde 193
Montezuma cypress 96
Mountain Home Grove 29–32
mountain pine beetle 178
mycorrhizae 176

IN THE CIRCLE OF ANCIENT TREES

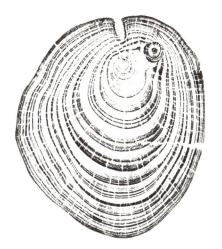

ACKNOWLEDGMENTS

This book came together as a collaboration between a proactive publisher, an editor with a vision, ten motivated authors with remarkable stories to share, and one incredibly talented artist. We thank the whole UniPress team for guiding us so expertly through the book-publishing process. We are especially grateful to Jason Hook, Daniel Mills, Kathleen Steeden, Wayne Blades, Slav Todorov, Richard Webb, and Jenny Manstead. We thank Blaze Cyan for her beautiful wood engravings and her dedication to adding an artistic dimension to this book. We also thank Greystone Books for publishing in North America—in particular, publisher Jen Gauthier and editor James Penco, whose helpful advice greatly improved the book.

Valerie thanks the authors for embracing her invite and suggestions and for sharing their unique stories. She thanks her colleagues at the Belgian Climate Centre and at the LTRR for their patience and perseverance and Wil Peterson for his support. Tom Swetnam thanks Chris Baisan, Tony Caprio, Peter Brown, Ramzi Touchan, and Linda Mutch for contributions in the collection, dating, and analyses of giant sequoia tree rings and fire scars, and Nate Stephenson and Dave Parsons for guidance and support. Momchil Panayotov is grateful to people who helped in research efforts of *Pinus heldreichii* forests in Bulgaria: Nickolay Tsvetanov, Velislava Todorova, Violeta Kotova, Albena Ivanova, Pepa Vasileva, Anita Kostadinova, Yanitsa Karamankova, and Martina Grigorova. Edmund February thanks Cape Nature for giving permission to work in the Cederberg and specifically on cedars, Peter Linder for interesting cedar conversation, Dawie Burgher for his enthusiasm, and Nicky Allsopp for her support. Gretel Boswijk thanks all those colleagues who have shared her research journey with kauri, especially Anthony Fowler, Alan Hogg, Neil Loader, and John Ogden. Special thanks to her family and those who left too soon. Matthew Therrell thanks his students, Rodolfo Acuña-Soto, Jose Villanueva Díaz, Matthew Gage, David Stahle, and public and private landowners in the United States and Mexico. Marta Domínguez Delmás thanks the editor for the invitation; Ute Sass-Klaassen, Elsemieke Hanraets, and Reyes Alejano for their guidance and inspiration; Naturalis Biodiversity Center and the Cultural Heritage Agency of the Netherlands for their support; and her loving family for everything. Daniela Granato-Souza thanks David Stahle for guiding her throughout her journey of learning dendrochronology and Ana Carolina Barbosa, Evandro Dalmaso, and Eliane Dalmaso for supporting field collections and allowing the development of our first tree-ring chronologies in the Amazon. Bao Yang thanks Peng Zhang, Yesi Zhao, Jingjing Liu, Chun Qin, Xiaofeng Wang, Minhui He, Xiaomei Peng, and Shuangjuan Wang for their assistance. Malcolm Hughes thanks the late Tom Harlan and Donald Graybill for their massive contributions to the data he discusses; and Matthew Salzer, Andy Bunn, Gary Funkhouser, Nicholas Graham, and Connie Millar for profound advances in the science of bristlecone pine. Jonathan Barichivich-Henríquez thanks the editor for her invitation, Antonio Lara for studying and defending the alerce forests, Nancy Henriquez for protecting the land and the Ngens, and Isabel Dorado for her support and discussions during a challenging writing time.

We are grateful for the funding for our research and for this book provided by the U.S. National Science Foundation, the National Natural Science Foundation of China, the European Research Council (ERC Starting Grants WoodCulture, Grant No. 101165305, and CATES, Grant No. 101043214), the National Science Fund of Bulgaria, The Water Research Commission, The Mellon Foundation of New York, the South African National Research Foundation, the (former) Foundation for Research, Science and Technology Tūāpapa Rangahau Pūtaiao, the Marsden Fund Te Pūtea Rangahau a Marsden, the Brazilian Federal Agency for Support and Evaluation of Graduate Education, the U.S. National Oceanic and Atmospheric Administration (NOAA), and the University of Arizona.

CONTRIBUTORS

Valerie Trouet is a Professor in the Laboratory of Tree-Ring Research (LTRR) at the University of Arizona. She is a dendroclimatologist, using tree rings to study the influence of past climate on ecosystems and human systems. She is the author of *Tree Story* (Johns Hopkins).

Thomas W. Swetnam studies forests, climate, and human history using dendrochronology. He is Regents' Professor and Director Emeritus at the Laboratory of Tree-Ring Research, University of Arizona. He has investigated wildfire history in the western United States, Mexico, South America, and Siberia, Russia.

Momchil Panayotov is an Associate Professor at the University of Forestry in Sofia, Bulgaria. He teaches dendrology and related disciplines. His research interests are focused on forest ecology, natural disturbance regimes, and tree response to climate variation and extreme events. In his studies he relies heavily on dendrochronology to obtain information from trees.

Edmund February has a Masters in Archaeology, a PhD in Botany, and a lifetime commitment to the Clanwilliam cedar. Now retired as a Professor of Biology at the University of Cape Town, he still regularly travels to the Cederberg to visit the cedars.

Gretel Boswijk is an Associate Professor at Te Kura Mātai Taiao School of Environment, Waipapa Taumata Rau University of Auckland. She studied archaeology, teaches historical geography, and has been involved in kauri tree-ring research since late 1999.

Matthew Therrell is a Professor of Geography at the University of Alabama. His primary research focus is on using tree-ring records to help understand how past climate variability has affected society and the environment.

Marta Domínguez-Delmás is a Senior Researcher at Naturalis Biodiversity Center and the Cultural Heritage Agency of the Netherlands. She has over 20 years of experience studying wood from material heritage to unlock the cultural, technological, and environmental information it contains, and she is still passionate about it.

Daniela Granato-Souza specializes in dendrochronology in the areas of dendroclimatology and dendroecology in tropical and temperate forests, with a focus on the Amazon, developing tree-ring chronologies from remote areas to reconstruct past climate variability, and understanding stand dynamics in the tropical forest.

Bao Yang is Professor of Physical Geography at the School of Geography and Ocean Science, Nanjing University. He utilizes tree-ring data in conjunction with other proxies to understand Holocene climate and environmental changes, and explores their impacts on both ecological systems and societal development.

Feng Wang is a Postdoctoral Researcher in the Department of Geographical and Sustainability Sciences at the University of Iowa. His research integrates tree-ring data to reconstruct Common Era climate patterns and investigate climate impacts on forest ecosystems across regional to hemispheric scales.

Malcolm K. Hughes (Regents' Professor Emeritus, University of Arizona) is a British ecologist who moved to Arizona in 1986 to direct the Laboratory of Tree-Ring Research. For 55 years he has focused on past climate as revealed by multiple properties of tree rings in many countries.

Jonathan Barichivich Henríquez is a Senior Scientist at CNRS (France) specializing in climate and global ecology. He combines tree-ring data from ancient trees with ecosystem observations and numerical models to study how global change affects terrestrial ecosystems, from tropical forests to boreal regions.